CW01657534

Love to All

From Wilfred Perera

BOAT IN THE BLUE

The wartime story of an RAF Air Sea Rescue crew and their boats

W. D. Pereira

Line One publishing ltd

Published by
Line One Publishing Limited
Bayshill House
Bayshill Road
Cheltenham
Glos.
GL50 3AA

ISBN 0 907036 22 8

We would like to express our appreciation to:—

The staff of the R.A.F. Museum, Hendon, who supplied photographs and from whose records the listings in Appendix B were extracted.

Group Captain L.R. Flower MBE.MM. for photographs and the foreword.

John Pritchard for his help and the use of his photographic collection.

CONTENTS

FOREWORD

by

Group Captain L.R. Flower MBE, MM.

Among the many people I talked with during my last official visit to RAF Mount Batten prior to my retirement in December 1970, was a young airman who had then only recently qualified as a Motor Boat Crew. I had earlier been briefed by his Squadron Commander that this lad had no less than nine 'O' levels and that he had passed his initial trade training course with exceptionally good results. I was more than a little impressed and I asked the airman why, in view of unusually good academic achievements, he had not opted for aircrew training or for one of the many ground trades with much greater technical content and career status than that of Motor Boat Crew? He answered that it was because he wanted to be a sailor. "Why then," I asked, had he not joined the Royal Navy? "Because", he said, "I wanted to join the Royal Air Force". To this I had no answer. Nor, I suppose, did I really need one. I had heard it all before. And after all, where else, over the years from 1918 to date, would we have recruited our RAF Marine Crews except from among those contrary characters who were determined to do their seagoing in light blue uniforms?

At one time there were as many as four thousand Royal Air Force sailors and the author and I were two of them. We served together in Air Sea Rescue during the last months of the war and in July 1945 we were on our way to the Far East in a Long Range Rescue Craft. Had we reached our destination it would certainly have confirmed our status as sailor men of some substance, quite irrespective of the colour of our uniforms. But when Japan unexpectedly surrendered we were detained at Gibraltar for four months awaiting orders. However, if there is one good thing one can say in favour of four months spent mostly in propping up the dockyard wall, it is that it provides an excellent opportunity for observing one's fellow crew members at very close quarters. In retrospect I do not think that I made good use of this opportunity but BOAT IN THE BLUE testifies that the author certainly did.

After 1st April 1986, however, there will no longer be a place within the Royal Air Force for those wishing to be sailors in light blue. For on that date the Marine Branch is formally to be disbanded and the long history of RAF marine craft will have come to an end. A sad thought for those who have spent part, and in some case the greater part, of their lives as RAF Sailors and not least, perhaps, for those who served in wartime Air Sea Rescue. In the circumstances, the publication of BOAT IN THE BLUE at this particular time could not be more opportune, and I welcome its contribution towards keeping old memories alive.

63ft high speed rescue launch with three Napier Sea Lion engines in Dover Strait area during wartime. *(W.D. Pereira collection)*

Chapter 1

UNDER TRAINING

The RAF selection board consisted of three officers. In the centre sat a flight lieutenant with one arm missing. On his right was a pilot officer with a burnt face. To the left a tough looking warrant officer stared hard at me for the entire interview. Probably I was the thirtieth they had seen that day. We were arriving at regular intervals for our medical examinations, then a long wait in an equally long corridor. Outside the air raid sirens sounded, followed by the bark of guns and the crump of bombs. For this was wartime London which the Germans bombed from beginning to end.

The flight lieutenant studied my file while I sat on a hard backed chair. There was rustling as papers were passed to other officers and handed back again. Heads came together for a whispered conversation followed by much nodding. I became aware of a clock ticking behind me. Eventually the flight lieutenant asked the first question of which friends had warned me and for which I was prepared.

"Why do you wish to join the Royal Air Force?"

The officers listened to my well rehearsed answer with impassive faces. Finally I could think of no more to say and there followed an awkward silence. Eventually the pilot officer seemed to rouse himself from distant thoughts. He asked me the second question.

"What would you like to do in the Royal Air Force?"

In those days, every lad of eighteen wanted to be a fighter pilot. While part of me cherished the same ambition, I knew it could not be fulfilled due to imperfect eyesight. When I mentioned my eyes, there was more nodding at the papers and in a way I was relieved. I answered that I would welcome whatever job they gave me, though I should prefer one which did not entail killing people.

The three officers gazed unbelievingly, I imagined, at me. So intense were their stares, I began to have visions of guards being called and being dragged to the Tower as some sort of traitor. In my mind's eye, I was already in front of the firing squad when the warrant officer asked me the third and completely unexpected question. His query seemed irrelevant yet was dead on target.

"How would you find the Pole Star?"

I experienced several seconds of inward panic before admitting I did not know. To my surprise the flight lieutenant appeared to approve of such ignorance. He said,

"Good, good. The Royal Air Force prefers to train people properly, instead of having to counter pre-conceived and generally wrong ideas. Very good. You'll be ideal, I think." There were nods from both sides. "As we all think. Excellent!"

For my part, I was bewildered. To me, finding the pole or any other star could only mean navigation. Was I to be air crew despite my eyesight? It was one matter to dream of glory, but quite another to be given the chance. I sat tensely while further papers were consulted. The pilot officer suddenly began scribbling away.

"Here," he said, holding out a form. "You have been accepted. And here," another piece of paper came my way. "This is your railway warrant."

Somehow I found it hard to take them. Already I was in a bomber crossing the enemy coastline with flak and fighters hurtling towards me. Such perils started in quiet rooms.

"Any questions?" the flight lieutenant enquired.

"Well. . ." I tried to think what to ask first.

"We felt," he went on, "as you did not want to kill anyone, it would be an ideal choice. Fortunately we have a few vacancies at the moment in that branch — which was why you were tested about finding the Pole Star. Any more questions?"

"Er, could you tell me the branch, sir?"

"Why, the RAF Air/Sea Rescue Service of course," he informed me rather sharply.

The way home from Kingsway to Wimbledon remains in my memory. While walking across Waterloo Bridge, I reflected on what lay in store for me. My world was changing as happened to everyone else in those eventful days. Graceful vapour trails of desperate aerial battles streaked the London sky. Occasionally I heard the faint chatter of machine guns which meant bullets tearing into metal and flesh. Yet trains still ran from Waterloo Station despite last night's bombing intermittently continuing through the day. The train took me past Vauxhall, Clapham Junction and Earlsfield — dismal districts at the best of times. That afternoon, I recall, they were awful.

Evidence lay everywhere of the constant bombing to which London was subjected. From the train one could see roads roped-off, fire engines and ambulances lined up at intersections, air raid and first aid workers scrabbling in the rubble of what had once been family homes. Gazing along the rows of terrace houses, those that had been hit looked like missing teeth. Many were smouldering and some still burning fiercely. As I watched, a wall collapsed. What harm it did I will never know. It was a typical wartime incident seen from a train miraculously threading its way through the devastation.

On arrival at Wimbledon, I learned from the ticket collector of several bombs hitting the district earlier in the afternoon. Next I found that my way home would have to circuitous. There were more barriers, more fire engines and ambulances, more devastation. At one place the road looked as if it had been chopped up by giant axes and burning gas flared between the slabs. I began to wonder if my home was still there. The stick of bombs seemed to have fallen in that direction. Two houses were gone from the road next to ours and another demolished at the corner.

To my relief, our house appeared intact. A closer view showed the windows, which were criss-crossed with tape, to be unbroken. At least, I reflected, my last night would be a comfortable one. My mother had been killed in the bombing. Following the funeral, my father went to stay with friends in the country. For the last few weeks I was alone in what had once been a happy home.

As soon as I reached the front gate, it became evident something was wrong in the house. Water flowing from under the front door had flooded the garden. On opening the door I saw the ground floor was also flooded, while more water cascaded down the stairs. I had previously learned how blast could play peculiar tricks and this was another of them. Although not breaking a single window, a pressure wave from the latest bomb had sheared the ballcock arm within our attic tank. Mains water then quickly filled and overflowed from the tank to create the situation which confronted me.

So my first job was to turn off the mains tap. After that came mopping up operations which took most of the evening. In this I was helped by friends and neighbours. Eventually I was alone in the house, left to my own thoughts. I could not help thinking how our seemingly secure family life had gone forever. It seemed as if parties, birthdays and Christmases never existed. There was nothing left but to go to bed and next day join the RAF.

The night was strangely quiet but morning started with a bang. A landmine had fallen near Wimbledon Station and many streets were closed to pedestrians. As I picked my way round the district, ambulances came and went from a shattered block of flats. An air raid warden with a grey haggard face told me it was bad. He experienced great difficulty putting what had happened into words. At last he blurted out,

"It's the glass. Some of the people are like porcupines."

Once more miraculously, the trains were running. I travelled back to Waterloo, then crossed London for Kings Cross and Bedford. There camouflaged lorries waited to take us to RAF Cardington. What happened next seemed like utter confusion at the time. In retrospect I have come to realise the place must have been well organised. Thousands of us were medically re-examined and innoculated, kitted out and inspected, drilled and instructed. Everyone who joined the air force went through this, so I shall pass on to the air/sea rescue part. Suddenly the tide of men began to divide into different streams. I was sent to a real air/sea rescue skipper who told me what would happen before I joined an actual crew.

"To begin with," I was informed, "you will do your motor boat crew training which takes three months. If you pass, then you move up from group five to three level of pay. You will also have a choice of remaining a motor boat crewman, with a chance of later becoming a coxswain. Alternatively you might like to try for a group one technical trade such as wireless operator or marine engineer. That would entail your going elsewhere on a second course lasting up to a year. You need not make a decision now, but should do so during your basic motor boat crew training.

"Motor boat crews are trained at RAF Corsewall, near Stranraer in Scotland. Everyone engaged in air/sea rescue has to go through the marine training school at least once. If you opt to remain at motor boat crew level, you will return to Corsewall from time to time for additional tuition. Technical training is held elsewhere. Marine engineers, for example, are trained at RAF Locking, near Weston-super-Mare. So you see, it is a long pull before one joins a crew."

At last my posting to RAF Corsewall arrived, together with a railway warrant that would take me from Euston to Stranraer. I recall being told the train left Euston at ten every night and reached Stranraer early next morning. It was to be my first long train journey under wartime conditions and I did not know what to expect. At least, by being early, I managed to get a corner seat. This was fortunate because every coach was packed to capacity as the hour of departure approached. Soldiers, sailors and airmen, all with their kit and many with rifles, kept arriving. The compartment seats were soon full, the spaces in between were also occupied, then the corridors were crammed until they were practically non-negotiable.

Luckily our compartment was occupied by RAF personnel because soldiers carried so much equipment while every sailor seemed to be drunk. My travelling companions were old hands at this game. The two airmen nearest the sliding door wedged it against intruders and got out playing cards. A Waaf said she was a cook and to prove it produced a large box of corned beef sandwiches. At midnight the two electric light bulbs were removed from the compartment ceiling and we settled down to sleep, another Waaf on my right immediately snuggling up against me.

We dozed rather than slept. The train alternated between racing across country or jerking through sidings and stations. At some stations there was a changeover of servicemen though the total increased. I remember looking out at one station — it may well have been inevitable Crewe — to see the platforms filled with thousands of officers and men. They were fully armed and alert in the early hours of the morning. Our own compartment enjoyed relative peace apart from one disruption. The Waaf cook went to the toilet and on the way back was set upon by the sailors. When we rescued her, she was half indignant and half pleased. Then we settled down again.

I was in a deep sleep when a red-haired airman who sat opposite shook me awake. He had raised the blackout blind allowing the first faint light of day to enter the compartment. The airman gestured for me to do the same which I managed with difficulty, being somewhat encumbered by a gently snoring Waaf. Outside were bare hills with grey clouds pressing heavily down on them and sleety rain blown horizontally across the bleak landscape. Lowly farmsteads seemed to be cowering beneath the elements. The airman however was strangely elated and explained the reason in two words.

"Bonnie Scotland!" he said, his eyes shining.

Presently, the rest of the compartment stirred. We men yawned, stretched and felt our stubbly chins. The Waafs decided that, sailors or no sailors, they had to go and freshen up. Meanwhile outside, Bonnie Scotland was substituting sleety rain for driving snow. It was the only choice of weather we would be given for the next three months. Even the train was reluctant to enter Stranraer Station. At last it came to a final juddering halt. No sooner were the doors and windows opened, when the night's fug was replaced by salty sea air. Most of the passengers headed for the ferry to Northern Ireland while waiting by the station were two RAF trucks. We set off along roads glistening with moisture. The very atmosphere seemed to be composed of water. Someone remarked on this and the red-haired airman proudly informed us it was Scotch mist.

Presently the road ran beside Loch Ryan and Wig Bay. After that we came to Kirkcolm village which was then two short rows of one-storey cottages facing each other. A small store and a wee kirk completed the community. To the right, a driveway led through the woods of a country estate. The trucks turned onto this drive and, as we bumped along it, more and more Nissen huts came into view. They had corrugated iron rooves with rough wood walls at each end containing a central door and two windows. All were painted in drab camouflage colours and sited under the ever-dripping trees.

Somewhere near the centre of the woods, there was a clearing round which stood larger-type Nissen huts. These were occupied by station headquarters, the camp barber, cookhouses and an assembly hall. No sooner had our trucks stopped outside the orderly room of RAF Corsewall than we were lined up for inspection by the station warrant officer. This was our introduction to Soapy. The nickname was derived from the way he obsequiously rubbed his hands together when talking to officers. With lower ranks however, his fingers kept curling and uncurling as if anxious to tear out throats, also froth would form at the corners of Soapy's mouth.

We were, Soapy told us, the scruffiest bunch of airmen he had ever seen. Our uniforms were creased, our faces unshaven and — worst horror of all — we had too much hair on our heads. Before being given breakfast and allocated huts, newcomers must have their locks shorn. Thus worn out from the night's travelling, followed by a wearisome journey in the backs of trucks, we had to queue outside the barber's shop. The barber was a Cockney with a keen sense of humour, honed no doubt by the constant stream of customers. No sooner had he finished with me and I emerged from the hut, when Soapy pounced again.

"Get your hair cut!" the station warrant officer shrieked as though I was a hundred miles away, his face mottling purple and red with outrage.

"I've just had it cut, sir."

My rash reply nearly caused Soapy to suffer a stroke. He began to froth freely which damped down further words. Instead I was waved to the back of the queue and eventually the camp barber welcomed me for a second session in his chair. Only after that thorough shearing, with the Scottish wind playing round our short back and sides, were we given food and shelter. Each hut contained twenty beds, ten per side separated by lockers. In the middle stood an iron stove flanked by two trestle tables and four benches. RAF Corsewall clearly believed in the spartan life.

I soon discovered that the marine training school included an amazing cross-section of characters. They ranged from amateur yachtsmen and professional fishermen to horny handed dockers and bargees. Among them was a sprinkling of those completely ignorant about the ways of the sea — such as myself. The inhabitants of our hut typified the mixture. My bed stood between one occupied by an Oxford man and another by someone from Cambridge. Both were on skipper's courses. The Oxford man was an amiable chinless wonder renowned for total impractability. It was said that if asked to knock in a nail, he would place the flat head on the wall then hammer away at the point. The Cambridge man was most charming, well connected and a rabid communist. He kept trying to lend me light little tracts by Marx and Engels. I do not know what happened to him. He would have been more at home as a senior mole in the Foreign Office.

But air/sea rescue was our reason for being there and the newcomers' course commenced early next morning. Like every RAF course, it was highly concentrated and applicable, also allowed no margin for slacking. Failure of any intermediate test, let alone the final examination, meant instant dismissal from the school and reversion to dreaded general duties.

The actual school was about three-quarters of a mile from our camp in the woods. There, either beside a pier or moored at buoys, the RAF had assembled the various marine craft with which we had to become accustomed. Starting at the lower end of the size scale were single-engine, 18 foot long planing dinghies and 24 foot marine tenders. Then came twin-engined seaplane tenders, which were about 40 ft long with a wheelhouse and passenger cabin. Next one graduated to 60 ft seagoing pinnaces having more powerful engines and crew quarters below decks. At the top end of the range stood the high speed air/sea rescue launches, though we were not allowed anywhere near them for many a day.

The first practical lesson we received was on how to row a boat. This proved fairly easy, then came the difficult bit. One oar was taken away and we had to scull from the pier to a raft — in a straight line. It was a knack which many took a long time to acquire. They were therefore obliged to keep trying, hour after hour, despite the unremittingly inclement weather. For the particularly inept, wooden fittings similar to the notched transom of a boat had been erected round the raft. Duffers and malingerers were put on the open raft with oars for 'sculling practice'. We would look sympathetically across the water, more often than not through the sleaty snow, to those unfortunates working away yet getting nowhere.

Only after passing rowing and sculling tests were we transferred to planing dinghies and marine tenders. Here the art was to bring craft alongside the pier or another boat without a disastrous crunch, which drove our instructors beserk. At the same time we learned to handle boathooks and heaving lines, fenders and moorings in the cold, wet conditions. I well recall slipping off an icy pier step and falling several sickening feet on to the edge of an engine cover. While wondering if I was in one piece, an instructor called to me.

"Why are you lying there? This isn't a holiday camp."

None of us were under any illusions on that score. After a morning on the boats, an afternoon in a classroom or workshop, then an evening studying, we were exhausted both physically and mentally. Nevertheless there were plenty of other duties such as preparing for parades, kit inspections and acting as guards around the camp. For the latter, we spent two hours during the night patrolling as far as the pierhead. After a couple of hours in the cold, it took as long again to get warm. By that time it was morning and the round of work and studying began once more.

The main problem at RAF Corsewall was trying to find some form of relaxation. During the evening we could take a break by visiting a canteen, but that was the extent of the camp's amenities. As we worked a five and a half day week, also as the Scottish sabbath was not conducive to having fun, this left Saturday afternoon and evening to try to escape from the pressures. To do so, we boarded trucks taking us to the fleshpots of Stranraer. There we had the choice of canteens or pubs and, in the evening, the local hop which went on from eight until midnight.

Stranraer's dances seemed promising. The band played loud and clear, also the girls were attractive — especially after a week cooped up at Corsewall. In fact they were most friendly and it was a shame, when dancing with them, to keep watching the clock. This was because we airmen, like so many Cinderellas, had to be back in camp before midnight. To make sure we did so, the last truck left at eleven. It meant that the local lads, quietly drinking their beers with whisky chasers in nearby bars, would arrive and take over the girls after we had departed. Our journeys back to RAF Corsewall were therefore not happy ones.

The situation produced some interesting results. A certain character I remember pretended to hide behind a tree whenever he saw someone coming. He would then leap out, at the same time firing a shot from the tip of his outstretched forefinger. Another type joined in, though his act was to pull an invisible pin from an equally invisible bomb. He did this with his teeth before lobbing the play-act missile in the direction of his mock victim. Those long days and nights, weeks and months under the dripping trees certainly provided us with a winter of discontent.

The sequence which followed is a prime example of how an unnecessary state of affairs can get sillier. Within the camp, certain personnel were members of a sports and welfare committee. If one wanted to make a suggestion, this had to be done through a representative who raised the matter at the next meeting. An idea was thus put forward which seemed eminently reasonable. Nearby lay an old and neglected golf course that had once been part of the country estate. The sportsmen among us said they would be happy to mow and tidy up the course if the committee could get them a few old clubs and golf balls.

The request went through the committee, then was passed up the line. Rumour told of it moving from office to office at Corsewall before winging its way to somewhere called group. Hopes were high, that is until a special parade was called for us to hear group's verdict. Lined up in the usual bad weather, we waited for Soapy to appear. The first signs were clearly ominous. His fingers were curling and uncurling. After a long glare at us, he spoke with ill-concealed annoyance at our presumption and stupidity.

"I was against the whole idea," Soapy said, "from the start. You're all here to work, not to play with golf balls. Besides, golf is not a game for the likes of scruffy, long-haired airmen. Golf, if they want to play it, is for proper commissioned officers — when they are not hunting, shooting and fishing. So group has confirmed what I thought. There will be no golf clubs, golf balls and you lot poncing about. Intead," and here froth began to increase, "you'll be expected to work harder. Also discipline has been getting too slack and I intend to tighten that up right away."

Right away meant we had to open our ranks for Soapy to inspect us. Half those on parade received taps on the shoulder and the curt command 'Haircut'. Others were booked for creased uniforms, muddy boots and dirty buttons.

On Sunday evenings, a film was shown in the camp assembly hall. Following the above mentioned parade, we were to watch a movie starring Alan Ladd and Veronica Lake. Alan Ladd, with his impassive face and laid-back manner, was supposed to depict toughness and meanness. Veronica Lake, whose long blonde hair kept falling over one eye, represented the zenith of wartime sexuality.

In the film, Alan Ladd was particularly tough and mean. At one stage, he kicked a kitten and at another thrust a crippled child in a wheelchair roughly out of his way. That was strong stuff for those more innocent days. Then came the most shocking scene — the ultimate. Veronica Lake was being difficult and Alan Ladd, without warning, knocked her senseless. There was utter silence in the camp cinema, before a voice remarked,

"I never knew Soapy had a son."

The laughter which followed practically raised the roof. Soapy wanted to stop the film show, but other NCO's prevailed on him to wait until afterwards. Afterwards we were paraded outside while Soapy ranted and raved. We knew the regime would worsen, yet not one regretted the incident.

What followed was a triumph of higher education over Soapy's relative ignorance. As I have related, we did have university men about the camp and some of them put their heads together for waging what they called psychological warfare. This commenced after church parade the following Sunday morning. Instead of everyone returning to their huts for a long and dreary day, we went to the disused golf course to witness what the brains among us had devised. Their scheme took the form of a golf tournament played without clubs and balls. The contestants were paired off and allocated caddies, then each player placed an imaginary ball on an invisible tee before driving along the tangled fairway.

I should relate it was an involved and exciting contest. Some claimed their ball had landed smack on the green. Others admitted to slicing into the rough. Players cursed and we cheered as they tried to extricate themselves from bunkers. There were amusing incidents and not a few surprises. One player near the hole missed an easy putt while another over thirty feet away succeeded to much applause. After an hour or so of this, we returned to camp in high spirits.

The exercise was discussed all week and repeated next Sunday morning. More people came to eye our eccentricities including villagers, aircrew from the seaplane base in Wig Bay and Soapy himself. He frothed a bit but said nothing. Then some officers arrived, nodded their heads and left. During the second week, many of the faint-hearted spoke about failure. However, on the third Sunday, a camouflaged staff car flying a pennant nosed its way through the crowd. An hour later there was a notice to the effect that golf clubs and balls were available in the sports and welfare office for those who wanted them. Back in our hut, my Oxford neighbour smilingly remarked on the walls of Jericho tumbling down and the Cambridge man claimed a victory over the forces of reaction.

Unfortunately the victors were unable to enjoy their spoils. By that time our course was nearing its end with final tests and examinations pressing hard upon us. I passed with my usual just above average, applied for a transfer to marine engineering and had it confirmed. We were delighted to leave the dripping woods of RAF Corsewall. Winter was coming to an end, but no one had told Scotland. On our way past Newton Stewart, Castle Douglas and Dumfries, the sleaty snow kept falling. Eventually, as the train curved southwards to the border, we could see blue sky over England and a bright celestial object someone identified as the sun. After that it was springtime all the way to London.

The train from Paddington to Weston-super-Mare was fairly full though not so crowded as the one from Scotland had been. The passengers included many civilians whom I later learned worked at the wartime establishments around Bath and Bristol. After Bristol, the coaches were virtually empty. I sat alone in a compartment — that is, until a Waaf arrived to join me. She was a trim, dark girl and, when passing along the corridor, had removed her hat. While shaking out her hair, the Waaf gave a glance to see whether the compartment was occupied, then stared hard at me. Next she turned, slid back the door and sat on the opposite seat with a peculiar smile.

We were soon in conversation. The Waaf said she was stationed at RAF Records, Gloucester and, as her parents lived in Clevedon, she was able to visit them every ten days or so. When I told her I was on my way to Locking for a marine engineer's course, she commented that I must have come from motor boat crew training at Corsewall. Before leaving the train at Yatton, the Waaf let me know her name which was Julia.

"You must visit us at Clevedon," she insisted. "I'll talk to my parents and write to you at Locking. That is if you want to. . .?"

Of course I did. Julia was good-looking and knew that part of the country. I was still thinking about her when the now practically deserted train ended its journey at Weston-super-Mare. From there, the few of us remaining were taken to RAF Locking situated about three miles south of the town on the far side of a small aerodrome. This new station turned out to be far different from Corsewall.

To begin with, Locking was a permanent air force base which, I believe, is still there today. The station consisted of four wings each with its own administration and training blocks, catering and medical facilities. Each wing also had a parade ground to which we were summoned by a band every morning for hoisting the RAF flag. That might sound corny nowadays but the scene sticks in my memory — the still lines of blue-clad figures, the band playing and the ensign fluttering in the spring breeze.

At the end of the parade, we went straight into a nearby training school. The marine engineering course of nine months duration was broken up every two or three weeks by practical tests and written examinations. It was the usual RAF pattern of concentrated instruction, though with a different approach when compared with Corsewall. Our instructors were engineers rather than seamen. If anyone lagged on a particular subject, they gave extra tuition — not as a punishment, but to help that person over a bad patch. Thus we were guided through basic metallurgy and the theory of internal combustion engines to practical working on various power plants. In addition to work, there were excellent sports facilities. Spring was with us, the town of Weston and the whole of Somerset there to enjoy. Also, within the first week, I received a letter from Julia inviting me to visit her at Clevedon.

A Locking truck, then a country bus from Weston, took me to Clevedon further up the coast. The town was much smaller in those days and seemingly untouched by war. I followed the directions Julia had given me. They took me to a substantial stone built house on a lawn high above the road. A curving drive led to a large turning space beside the garage, greenhouses and front door. Julia's parents were clearly people of substance. There was a big car in the drive, a gardener tended the flowerbeds and a uniformed maid opened the door. Most other people I knew had lost such luxuries since the outbreak of war. For some unknown reason, I became increasingly apprehensive.

Julia was the first to greet me. I was shown into a large drawing room with a bay window overlooking the lawn as well as the town. The whole place, furnishings and decorations, were old-fashioned. Heavy draped curtains flanked the window, crystal ornaments lined the marble mantlepiece and there were several rows of silver-framed photographs on top of the half-closed grand piano. Although the room had a sizeable floor area, one had to be careful when moving about it due to innumerable statuettes, floral arrangements and slender occasional tables.

Presently Julia's parents arrived. I found them pleasant and polite, though with a peculiar air of reserve. The father was short and dapper, the mother tall and watchful. Both had sad, strained faces. The most disconcerting part was the way they and Julia kept staring at me. Something was obviously wrong, but I did not know what.

After tea, Julia took me for a walk round Clevedon, also to see the front. We were met by several other strollers who all stared hard at me. Later, when we returned to the house, it was suggested that I should stay for supper. They kept insisting and Julia's father even said he would run me back to Locking. I declined, so as not to outstay my welcome, yet I increasingly felt there was something not quite right, some sort of missing factor. As I boarded the bus, Julia said,

"I do hope you'll come again. We so much enjoyed your visit."

The marine engineering course continued and I went to see Julia two or three times more. We were then what was called walking out but in a curiously circumspect way. If I took Julia's hand she seemed pleased though anything else was not exactly encouraged. One Sunday we decided to walk to the end of Sand Point overlooking the Bristol Channel. It was a superb day and naturally, after such a long outward walk, we found a resting place in the sun but out of the wind. Julia permitted a few kisses before sitting up with her arms round her knees. At a loss what to say next, I asked about her father.

"Oh, he does most things," Julia replied vaguely. "Property, commerce, various business interests. He's on several committees and a councillor, of course."

That seemed to represent the extent of my knowledge about Julia and her family when suddenly all was revealed. Following our day on Sand Point, we returned to Clevedon. Julia went upstairs to change leaving me alone in the drawing room. Idly I began to study the rows of silver-framed photographs on the piano and, in the very centre, was what looked like my picture.

A closer study revealed that while the person ressembled me, there were differences. The hair was not quite so dark as mine, the face was thinner and the airman wore pilot's wings which I certainly did not possess. Apart from these minor variations, the young man in the photograph was almost my double. The moment Julia entered the room I asked her about him.

"I should have told you before," she said simply. "That is a picture of my brother who was killed in the Battle of Britain. He was the apple of my parents eye as well as my favourite person. When I first saw you on the train, I had to get to know you. To ask you here. But I did not know how to tell you. I'm afraid we've played it all wrong, I'm so sorry. . ."

I did not know what to say, nor do I recall how we got through the tea and the rest of that day. On our way to the bus stop, I made useless remarks about being busy on my course, while Julia pleaded with me to visit them again. She wrote a couple of times from Gloucester, telling me when she would next be at Clevedon, but I just could not manage a reply. Ours had been a strange relationship and, without my realising it, I almost immediately became involved in an even stranger one.

As spring turned to summer, I decided to have my bicycle at the camp. Many people there had cycles, either for riding to and from nearby homes or to get away at weekends. In the first instance however, one had to obtain permission. This meant a visit to the wing orderly room beside the training school. A pleasant Waaf at a desk by the door gave me a form and, when I had completed it, she took the piece of paper to a warrant officer half hidden behind a line of filing cabinets. He peered over the cabinets at me, signed the form and that cleared the first hurdle. A phone call to my father disposed of the next. He put my bike on the train and, because there was no computerised super-efficiency in those days, it arrived the following morning.

The same evening, I cycled into Weston, returning via Brean and Bleadon. I had a wonderful feeling of freedom, not that life at RAF Locking was onerous. Trying to obtain a map in wartime was more difficult. At last a colleague lent me one and from then on there was no holding me. Every evening I went for a ride, either taking along the work I had to do for the next day or returning to it with renewed vigour. At the weekends I went further and further afield.

It was on one such weekend I met Hilary. She was the pleasant Waaf from the wing orderly room who had handed me the bicycle application form in the first place. Hilary had light gold hair, pale green eyes and a dusting of freckles across the bridge of her nose. She was a tall slim girl, the picture of health. I was on my way through Winscombe to Cheddar, had stopped at a little shop-cum-cafe for a drink and saw her sitting alone at the window table. It turned out she too was going to Cheddar and subsequently we cycled there together.

During wartime there was little or no traffic on the main roads, which meant our subsidiary lane was practically deserted. Hilary and I rode side by side, chatting as if we had known each other all our lives. She had always liked cycling and explored the surrounding countryside every weekend. I said it was my intention to do the same and by the end of the day we agreed to meet again on similar ventures.

The following Sunday we went beyond Cheddar to Wells. Hilary had been there before and was pleased to show me around. The tiny town nestling under the Mendips, the great cathedral, the bishop's palace built like a miniature castle complete with moat, fascinated me. I was also becoming more and more fascinated by Hilary. What had I done to be rewarded with such an attractive girl for my sole companion? She was knowledgeable as well. After showing me where swans rang a bell to get food, Hilary took me into the cathedral to see the animated clock which she told me was six hundred years old.

"A monk from Glastonbury," she explained, "called Peter Lightfoot invented it. Apart from telling the time, his clock gives phases of the moon and positions of the planets. That figure there, sitting with his heels against the two bells, is Jack Blandiver. He kicks the bells with his heels, then four mounted knights gallop round and round with one of them being knocked over at each circuit."

We had a magical day and we managed to repeat it for several more weekends. Among other journeys, we rode over the Mendips as far as Midsomer Norton. We took a train, with our bikes on it, to Bridgwater, then explored the Quantocks. Once an RAF transport driver gave us a lift to Bath and, after spending the most part of the day there, cycled back through the summer evening to Locking. The driver said he was passing through Glastonbury the following Sunday and would again take us on the outward trip. It seemed as if my good fortune would continue forever.

There was just one little difficulty. Hilary and I were never more than friendly with each other, though I wished to alter that. When Hilary told me she knew of a lovely but lonely ride back from Glastonbury across the Somerset plain, I believed my opportunity was near.

The return journey was lonely as she promised. The flat plain made it perfect cycling country yet was never dull. Every few hundred yards we came to brooks, streams and huge man-made drainage ditches. It was then high summer and the fields full of wild flowers. Willow trees swayed over waterways and skylarks sang above us. At last we stopped by a picturesque bridge. I tried putting my arm round Hilary's waist, but immediately she stiffened.

"Please," she said, "Not that. There mustn't be that. We're such good friends."

I tried talking to her, pleading with her, to no avail. In the end we set off once more and, although the sun still shone, the day seemed duller. Silently we rode on and on through the villages of Wedmore, West Stoughton and a windmill by Chapel Allerton. At last we came to Bleadon Hill where we had to stop for a rest. When alighting, I saw that Hilary's eyes were full of tears. Suddenly she lay with her back to me sobbing. Nothing that I could do or say seemed to comfort her. The sobbing went on for a long time and finally ceased. Hilary then wiped her eyes and face, got on her bike and rode ahead of me. On reaching RAF Locking, she went straight into the Waafs quarters where I could not follow her. Something was desperately wrong and I did not understand any of it.

Bewildered by Hilary's behaviour, I decided to wait a few days before contacting her. It began to seem to me that while my colleagues enjoyed fantastic adventures with the ladies of Locking, Weston and the rest of Somerset, I was fated to quasiplatonic disasters. By Thursday however my resolve weakened sufficiently for me to visit the wing orderly room. Immediately on entering it I saw that Hilary's desk was clear and unoccupied. A nearby clerk told me she had left. When I began to question him, the warrant officer rose from behind his row of filing cabinets and beckoned me.

I should mention that the Locking warrant officer was very different from Soapy at Corsewall. This one had his home in nearby Weston, two lads in the air force and he acted more as a fatherly figure towards us. For example, the Locking warrant officer was keen on sport, refereeing football matches during the winter and in the summer umpiring cricket. He repeated that Hilary had gone. When I again asked why, the warrant officer placed a gentle arm round my shoulder, at the same time saying,

"Come, come lad. You of all people should know the reason why. The lass is pregnant."

I became aware that the rest of the office had fallen silent. Not a paper rustled and the normal rattling of typewriters was stilled. Then, as if from a distance, I heard my own voice protesting ignorance and innocence. We had simply gone on cycle rides together. Nothing else. The warrant officer believed me and said I should use it as a lesson. There was my course to complete, a war to be won. I was in no position to become involved with the wiles of women. They were too tricky. They did not think like us men, he informed me — not even built like us. However he could give me her parents address in Cheshire if I wanted it.

As with Julia, I suddenly did not wish to pursue the Hilary matter. Too much lay ahead. At that stage of the war, the Germans were rampaging over most of Europe. Yet at RAF Locking, as at other training establishments, there was no shortening or simplifying of our courses. Standards were not lowered and we did not expect them to be. It sounds trite to say so now, but I do know from the first hand how — when in adversity — the British can exhibit a grim determination. That period so long ago represented the low spot of the war and the highest morale. Our examination marks were also high because we had worked hard. Thus we became air/sea rescue marine engineers, ready for our next posting.

As our stay at Locking neared an end, we were given a talk on the disposition of rescue bases around the coast of Britain, as well as in the Mediterranean, Middle and Far East. Afterwards we were handed forms on which to state where we wished to serve. I wrote Dover and a couple of days later learned I would be joining a high speed launch crew there who required a replacement engineer. I did not have the slightest idea what lay in store for me.

Chapter 2

DOVER DAYS

Perhaps it was my imagination, but the London to Dover train seemed to be behaving in a most peculiar manner. To begin with the train was reluctant to leave London and lingered through the interminable south-eastern suburbs. Then, once out in the open country, it attempted to break speed records making a blur of Kent, the Garden of England. This headlong rush ceased as we neared Dover and was followed by a series of long unexplained waits. Eventually the train jerked into life again and approached the station with a determined rush. As we slid beside the platform and came to a crashing halt, I heard voices shouting,

"Dover, Dover. This is Dover. All out, all out!"

My compartment was towards the rear of the train and I followed the rest of the passengers who seemed in an undue hurry to reach the ticket barrier. When I started after them, wondering why they were in such a rush, the train reversed out of the station and disappeared. I watched it go, turned once more towards the barrier and noted with alarm there was not a passenger in sight. Nor was the ticket collector visible. He had vanished like the rest.

Thus I walked straight out of the station and into the town to be confronted by similar inexplicable happenings. There were a few people about, but they too melted away on my approach. Some let themselves into houses and others darted up sideroads and alleyways. The place was rapidly becoming deserted. I began to feel like one of those Western bad men entering Laramie or Dodge City.

Before I could give more thought to this strange situation, there came a sound which can only be compared to an express train travelling through the air. The sudden screech ceased almost as soon as it started and was followed by a tremendous explosion. Looking in the general direction of these frightening noises, I saw a white cloud materialising against the blue sky while bits of Dover's famous cliff rained down in the vicinity. I was still absorbing this when a door by my elbow opened and a huge woman pulled me into her house saying,

"Jerry shells. Same time every day. Didn't anyone tell you?"

The hallway in which I found myself was reinforced with heavy baulks of timber which would withstand everything but a direct hit. Seated halfway under the stairs was a grizzled old man — her father as it transpired — reading Titbits magazine. Every time there was an explosion, and several followed, he became more immersed in the magazine. At last, after a long silence outside, the woman remarked to me,

"That's it then. No more for today. They use poor Dover for target practice regular as clockwork. Bloody Huns!"

We had a cup of tea, then the man said he would take me to the air/sea rescue base. He was not impressed with the RAF as sailors, being an ex-Royal Navy man himself. He informed me that airmen should have left such matters to the navy instead of playing at it. I was suitably deflated and yet at the same time interested in the way Dover had come to life again. People reappeared in the streets, the shops were already open and the old man kept pointing out landmarks to me. Only his landmarks were pubs and the information he gave me concerned their clientele. Apparently the navy had been going to the best places for generations and he warned against airmen trying to drink in any of them.

I was guided as far as the great gates at the entrance to Western Docks. There my papers were checked three times — by army sentries, then in a naval guardroom and finally at a wooden hut with an RAF roundel above the door. The entire area was bristling with boats. Moored beside the wall and out in the harbour were destroyers, corvettes, minesweepers, torpedo and gun boats. They were painted a dull grey and many looked considerably battered with obvious patches about their sides and superstructures. Tough looking sailors swarmed everywhere. An equally tough looking flight sergeant in the RAF hut said to me,

"Ah yes, you're the engineer replacement we requested. Where are you from?"

"Locking," I replied.

"No experience," he commented then added. "Well, you've come to the deep end. Off you go and jump in." I must have appeared a trifle apprehensive, for he suddenly smiled and went on to say, "Yours is the first launch along the dock."

Back at Corsewall, the rescue launch used for training was beautifully painted and polished. What I now saw were workaday boats, somewhat the worse for wear after months of arduous operations. Stains, rust and exhaust marks smudged decks and sides. Yet these craft looked right with their direction finding cones, crash nets and twin gun turrets. All else lay below.

Below was crammed to capacity with men and more equipment. The three powerful engines, which drove launches up to 35 Kts, occupied about a third of the interior space. Next, built in somehow, came a wireless cabin and a chart table beside the coxswain's wheel and compass. A hatchway and steps led steeply down into the bowels of the boat. There was a small sickbay for survivors and a microscopic cabin for the skipper followed by a rapidly narrowing foc'sle within the bows. The foc'sle had bench seats on each side, let-down bunks above them and a central table which was also of the let-down type. In tiny alcoves off this already small space was a galley with a paraffin stove for cooking and a triangular washing area with a toilet beside it.

Our nine-man crew comprised a commissioned officer as skipper, a first and second class coxswain, two deckhands, two engineers, a wireless operator and a medical orderly. The first essential — even before that of competency — was the ability to fit in and work together as a crew. We had been told this ad nauseum while under training and I therefore stepped aboard with some trepidation. What if my face did not fit? In order to try to obtain acceptance, I decided to say as little as possible, observing and listening instead. Fortunately this worked. It also left me with clear and indelible impressions of my air/sea rescue companions that have remained with me to this day.

The initial impression was of how the characters were dressed. No one appeared to be wearing RAF uniform as I knew it. Instead they were clad in an assortment of oilskins, sou'westers, rollneck pullovers, flak jackets and footwear ranging from running shoes to flying boots. There was no indication of rank and some days were to go by before I sorted everyone out. Meanwhile I was given a space beneath a seat in which to stow my belongings and shown the bunk I would occupy when it was lowered. It could not be lowered at the moment because four crew members were playing cards on the central table and two more preparing a meal beside them. So I sat and waited until one of the cooks said to me,

"Kettle's on. Care for some tea?" This made me feel slightly less invisible.

As I was sipping the tea, a huge man swung himself below instead of using the steps and came crashing down next to me. This turned out to be our skipper, a larger than life character six foot four inches tall and weighing some sixteen stone. He was dark haired and eyed with an urbane cast to his features. When he spoke, whatever the circumstances, it was always in an amused drawl. Skip, I later learned, came from an upper crust family with generations of wealth and estates behind him. In all the years I knew Skip, I never heard him give a direct order. Instead he would suggest it might be a good idea to do this or that, whereupon everyone fell over backwards to help him. Off the boat, we used to be rather proud and superior about our captain because of his top connections. His visitors were either very high ranking officers or gorgeous girls. He treated both as he did us with an air of amused tolerance.

Next in command came our first coxswain, an ex-trawlerman from Seahouses in Northumberland. Nicknamed Timber, he was another big man like the skipper though not quite so tall. Fair-haired and blue-eyed with a ruddy expression, Timber exemplified the British seamen who down history made a living off our coasts, when not following the Drake or Nelson in times of war. He invariably wore a spotlessly white rollneck pullover, also white seasocks turned down over the tops of his boots. If Timber had a fault, it was he never relaxed, especially once we had put out to sea. He kept working and he expected the rest of his landlubber crew to do the same despite weather conditions and bouts of seasickness. Every day Timber would write a long letter to his wife and he usually received one in return. When off-duty, he enjoyed a few glasses of beer or tots of whisky. Snooker was his preferred game and in fact the first question Timber addressed to me was whether I played.

"A bit," I replied cautiously.

"Good," Timber nodded and continued writing in his slow careful hand.

The second coxswain was called Fancy. He was the first to talk at length with me saying he had been a floor walker in a well known London West End store. Despite his mincing step, lisping voice and limp wrists Fancy was not — so another member of the crew later informed me — as queer as a nine bob note. I must admit being doubtful because Fancy used to give high shrieks when excited. He was of medium height and build but handsome as a matinee idol with deepset eyes. One could visualise him wooing the lady shoppers. Yet he was still a bachelor. Despite being totally unlike Timber, the two worked well together for reasons Fancy once explained to me.

"We have different talents, you see. Timber's a natural seaman while I possess a shopkeeper's mind and eye for detail. I see my job as having everything right for him to operate. That means equipment, stores, rations and the rest. When all's said and done, I wouldn't know one end of a marling spike from another. Or would I?" Fancy gave another of his high pitched shrieks.

"How did you get into air/sea rescue?" I asked.

"Press ganged!" he shrieked again. "Never in my life did it occur to me that I'd be put in a little boat with such an assortment of men."

We were indeed an assortment. I do not think the recruiters, planners and, nowadays, computers could have come up with a more varied selection. Immediately responsible to the coxswains were our two deckhands — Taff from the Valleys and Paddy from the Macgillycuddy's Reeks.

No one could help liking Taff. He was a simple, happy chap from a mining family. His father and elder brothers had all gone down the pit, but Taff wanted a different sort of life. He tried to join the merchant navy and ended up in air/sea rescue. Taff had a wonderful Welsh singing voice and whether he was swabbing the deck, manning a gun or cooking a meal, we would be treated to excerpts from hymns and operatic arias. Physically he was short, dark and hairy, which made many a girl say she would like to cuddle him.

Paddy was an altogether different proposition — lean, fair, full of charm but prone to sudden violence as we were to discover. He was also a bit of a mystery, because we never could find out what he had once done for a living. When asked a direct question, Paddy would go off on an Irish tangent. He was not so much evasive, his mind simply seemed to work another way. The skipper, who knew Paddy's record, once hinted at a chequered career. Paddy had joined the RAF at the outbreak of war and there was some talk of him training for aircrew. Of this, he made such comments as,

"Sure but you see they didn't take to me and I didn't take to them."

"There was a wee bit of a mix-up, you understand."

"It was as good as a bad play."

Paddy did not like the English and often, when we were out drinking together, someone would egg him on. It was funny then hearing him say how the IRA would sort us out one day. Yet he served us well through the war years and our only complaint was having to tie Paddy to his bunk when the drink was in him.

The second engineer and the person I worked closest with was Jock. He had been through Locking before me but passed at a lower grade, thus nominally I was in charge of him. His slower engineering progress was not surprising because in civilian life Jock had been a forester. He told me that, coming from Scotland, the RAF automatically assumed him to be an engineer.

"Sassenachs," Jock grinned, "think we all live round the Clyde."

"Where are you from?"

"North of Loch Ness," he answered. "Up in the empty bit." We had to go through a war together before I would appreciate the truth of his remark.

Jock had acquired a wealth of practical experience on boats while I was still training, so we worked as equal partners. Although seemingly solid and dour, Jock had a dry wit and, like most Scots I met, was extremely generous.

Sparks our wireless operator was in direct contrast to Jock the ex-forester. Here we had to contend with a sharp-eyed, quick-tongued Cockney up to every trick, particularly those connected with making women and money. Sparks had rackets running everywhere we went — clothing coupons, food rations, cigarettes and sweets. He really came into his own when the Americans arrived.

Such then was the crew on my arrival. We were short of a medical orderly but one was on his way. Meanwhile I began to find out about life in wartime Dover. I learned that until three weeks ago, it had been all hell let loose. Every night, the navy's light coastal forces were out battling with German E-boats. Every day, there were aerial dogfights over the sea with planes being shot down and air/sea rescue collecting the bits and pieces. Then came a curious but welcome lull, time to restock and overhaul. During it, two of the crew were taken off the launch. One was the engineer I replaced. Apparently he developed a stomach complaint due to high speed pounding on wave after wave. The last medical orderly had also gone because of incessant seasickness. Now two newcomers were to be tested — me and another orderly due to arrive in a couple of days.

"Do you get seasick?" Timber asked me.
"I might to begin with."
"Hm," he pursed his lips.

It took time and consideration to live in such cramped conditions. I thought this impossible the first night. At last there we were layered in close proximity, the skipper wedged in his slot of a cabin with the door left open. Paddy was a restless sleeper, Taff snored musically and Fancy gave little squeals when dreaming. There were plenty of other sounds — the perpetual lapping of water, our hull nudging the harbour wall, the measured steps of sentries and occasionally the sounds of distant engines. After sleeping, then dozing, I became aware of daylight through the portholes. Outside there was much activity, orders being shouted and a bugle sounding. All of a sudden Taff, who wore the trouser part of some old pyjamas, shot out of his sleeping bag and up through the hatchway. I listened to his bare feet padding across the deck and wondered.

"What was that about?" I asked on his return.

"Nearly forgot to hoist our flag." Taff was panting from his exertions. "Give me a hand to make breakfast."

Taff and I fried bacon as the others slept on. Presently people awakened, staggering in turn to the tiny toilet and washbasin. While we were eating breakfast, also reading the morning papers delivered personally by a newsagent friend of Sparks, the skipper announced we would be leaving harbour to give our engines a run. Jock had been working on them for the past week and, Skip added, it would offer me a chance to become acclimatised.

My main impression was the casual way everything took place. After a second cup of tea, Jock and I went aft to start and warm up the three Napier Sea Lion engines. It was deafening in the engine room and outside the deep growls from our exhausts echoed and re-echoed about the harbour. Then Taff went forward to cast off and Paddy came our way to do the same. Timber had signalled slow ahead from the wheelhouse and we were heading towards the harbour entrance. I put my head up through the hatch to see Paddy pull a Daily Mirror from his pocket, sit on the upturned dinghy and start to read.

Looking beyond him, I noticed a light flashing at us from Dover Castle. This gave us permission to leave harbour. The chalk cliffs, symbol of fortress England, were brilliant white in the morning sunshine. The sea outside was truly a surging green moat separating us from enemy-held France. We turned to starboard and headed for Dungeness, at that time an empty headland without its power station. Engine revs were progressively increased and the hard chine hull began to plane. Luckily for me the water was calm and I could enjoy rather than disgrace myself on my initial trip.

Near Dungeness we shut off engines and rolled gently in the lee of the land.

There was not another ship to be seen, not a single plane in the sky. The whole scene looked so peaceful it was hard to believe we were at war with anyone. I went forward to the wheelhouse and found everyone there staring intently at the smiling sea. Following their gaze I saw a loose mine drifting past us, the detonating horns clearly visible. Later in the war I would witness the havoc a mine could cause. Meanwhile, as we did not have to maintain radio silence, our skipper informed the navy of the mine and requested permission to blow it up by gunfire. No, the navy replied, they knew how to deal with such matters and would be out presently. They came three and a half hours later.

This was typical of the navy's superior attitude which we were to endure during the years that followed. Another example awaited us on our return to Dover Harbour. There pacing up and down the dockside was an irate naval lieutenant.

"The C in C wishes to see you," he curtly informed our skipper. "At once."

Skip was gone for the best part of an hour. On his return he looked rather bleak and asked if there was any tea going. The skipper sipped at it for a while before telling us what had happened.

"The admiral's a bit peeved. Hard to believe he's an old friend of my father. Not at all happy." Two more slow sips were taken. "You know how the navy likes to make a thing about hoisting flags? Well it seems this morning there was no one doing so on our boat. That is" another sip, "until someone in pyjama trousers rushed on deck in the middle of the ceremony." There was a choking sound from Taff, but the skipper ignored him. "Then," he went on, "you also know how the navy likes to dress ship when leaving harbour? One or two of the chaps up forward. It seems we had no one — apart from a fellow sitting on our dinghy reading a newspaper. No, the C in C was not pleased. Thought he was going to have a stroke. So Timber," Skip ended, "we'll have to show a bit more willing in future."

Nothing more was said about the incident. I was reflecting on how different Dover and the war was to my expectations, when there came a brief whistling sound followed by a loud explosion. A moment later we heard aircraft engines rev up as the sneak raider pulled away and headed for France. Several more incidents happened during the twilight hour. The Germans, I was told, were reverting to recent sneak raiding tactics. It meant we would not be spending the night moored by the open dock but would take shelter in the nearby submarine pen. This massive structure had been built to protect submarines, other naval craft and air/sea rescue boats in that order of priority. The building had room for about twenty-four vessels and was covered by some thirty feet of ferro-concrete.

We managed to find a space and, while manoeuvring into it, I viewed the structure with wonder and admiration. A submarine, two frigates, also rows of torpedo and gun boats lay quiet alongside a series of projecting walkways. At the inner ends of these, steps led to a crosswalk extending the whole width of the pen. Off the crosswalk were washing and toilet facilities for the crews. Green lights high overhead gave the vast interior a subterranean glow.

It was late and, while the rest of us were content to clean our teeth and splash some water over our faces before turning in, Fancy decided to take advantage of the submarine dock facilities by having a shower. From his locker he produced a floral towel and a cake of scented soap which caused some ribald remarks. Sparks warned him not to drop the soap as that would mean him bending down to retrieve it. Fancy took this in good part and left us getting into our sleeping bags. Just as I was falling asleep, he shook me awake saying,

"You must, you simply must come and see their toilets."

Fancy told me this was the first time anyone from our boat had gone upstairs in the building and what he had seen was a revelation. Eventually Sparks and I agreed to go with him. We reached the crosswalk off which were doors at regular intervals. These doors were labelled — ordinary seamen, able seamen, petty officers, chief petty officers and commissioned officers. We had noticed the labels on mooring up, but only Fancy discovered the differences. He told us that after a shower in the washroom nearest to his rank of corporal, he decided to see what the rest contained. Everyone had retired for the night and it seemed a good opportunity.

Starting with the ordinary seaman's room, Fancy noticed their toilets had no seats and were separated one from another by waist high partitions. The able seaman's lavatory bowls had two side pieces of wood riveted to the porcelain and the partitions were chest high though still without doors. Petty officers enjoyed plain wooden seats, normal height partitions and unlockable doors. Next came the polished seats and lockable enclosures for chief petty officers. Finally in the commissioned officers washroom there were sumptuous, spacious lavatories with a choice of toilet papers.

I did not think these differences so important but Fancy kept pointing out others and commenting on them. About the plain brass cold water taps in the seaman's washroom compared with chromium plated faucets for officers. About the lack of mirrors for other ranks while the commissioned enjoyed bevel edging. What concerned Fancy was that someone had thought up and applied the many differences — not only during wartime — but at England's front line town. After a while, Sparks began to argue with him.

"It's your old class system."

"My class system?" shrieked Fancy.

"Yes, you help it along poncing about West End stores pandering to women who have more money than sense. And where does that money come from? It comes from their men in the City who are in turn milking it from the rest of the country. An honest bloke could work all his life and not make what those city types pick up in an afternoon. They're the ones at the top of the class pyramid and you support them. The rest are squashed at the bottom, like ordinary seamen."

By then, the ordinary seamen were coming awake. Angry voices sounded below.

"What the hell's going on?"

"It's the bloody RAF."

"They need sorting out."

Somehow I managed to get Fancy and Sparks, still arguing about the English class system, back to our boat. Next day a fresh subject was raised. The crew began to speculate on the medical orderly due to join us. I learned that the last man had been persistently seasick. Timber then put forward a theory about medical people becoming so used to the unpleasant sides of life, they were more affected than most in other directions. In his experience, medical orderlies were the worst sufferers of seasickness among air/sea rescue crews. Taff took up the subject, talking about the Docs he had known. It seemed every orderly was called Doc and I idly wondered why our entire crew had a nickname but me. I was addressed or referred to as Wilf from the beginning to the end of my service days. Later, when I knew the skipper better, I remarked on this. He thought for a moment before saying,

"I suppose it's because your name is a rather quaint one. If I were you, I'd be grateful at not acquiring a nickname. At Eton, they called me Sticky."

We were nearing the end of breakfast and speculating on the new Doc, when footsteps sounded across the deck and down the ladder. To the surprise of everyone, there were two people — our replacement medical orderly and his wife. She was a uniformed nurse who became known as Mrs Doc. Apart from being male and female, Mr and Mrs Doc were almost identical in shape, size and looks. Both were short and slightly built. Each had benign expressions and wore small, round metal-framed glasses. They stood smiling at us and we stared at them.

While we continued to stare, for none of us had never heard of wives accompanying their husbands in such a way, Doc explained how his good lady wife had managed to get a nursing job in Dover. She had also found accommodation in town where they had stayed overnight. Doc went on to say Mrs Doc was off to work, but wished to see the boat he was joining. Fancy at once offered the lady a seat and fussed over her as if she were one of his West End customers. Mrs Doc responded by inviting him to share a meal with them at the earliest opportunity. Eventually she left and, after seeing her from the boat, Doc said to us,

"Marvellous woman. So resourceful. Being a nurse, she can pick and choose where she wants to work. And where better than near to me? Intends to keep me out of mischief she says, but that's only in fun. To tell the truth, she's very fond of her conjugals."

Conjugals or no conjugals, the war suddenly took priority. The arrival of our medical orderly coincided with the ending of the lull at Dover. All was activity. German sneak raiders began striking everywhere. Their aircraft and E-boats kept attacking our coastal freighters which had been slipping safely round south-eastern England for the past three weeks. The E-boats, operating mainly at night, proved the worse menace. These powerful diesel-engined vessels were heavily armed with cannon and torpedoes. They were also capable of carrying and laying mines in our sea lanes. Clearly they had to be stopped.

If I have been critical of naval stuffiness, now is the time to make amends. The Royal Navy was there to fight and that it undoubtedly did. As each afternoon wore on, RN ships became centres of feverish preparation. Great quantities of ammunition were taken aboard. Heavy machine guns and cannon were cleaned and loaded. Often, more armour plating and anti-shrapnel mats were installed. Crew members donned flak jackets and carefully settled into their action stations. They were under no illusion as to what lay ahead of them.

During the deepening dusk, one engine after another would be started until the whole harbour was charged with roaring, crackling life. Then boats began to move away from their berths to join the purposeful line of vessels heading for the harbour mouth. Their departure was achieved within a minute or two, after which only the engines could be heard from the outer darkness of the Dover Strait. The noise lingered for a much longer period, diminishing imperceptibly until replaced by the eternal sounds of wind and wave. The wind seemed to blow colder. Waves sucked hungrily at the harbour walls where so many vessels had been moored. A few officers and men lingered on the deserted quays before retiring to their base buildings to wait.

Waiting took most of the night. In the early hours of the morning, long before the fighting ships returned, a new kind of activity began. The first to arrive were the ambulances called blood wagons. They would line up along the dockside, then their crews would open the rear doors and set out medical equipment like merchants on market stalls. After them came the repair crews — engineers, armourers and salvage experts. They were joined by mobile cranes and trolleys filled with cutting tools. Last but not least were the naval surgeons and nurses. The surgeons were identifiable by thin bands of scarlet between gold braiding, the nurses by long capes which they wrapped round themselves against the cold, sobering hours before dawn.

At last, when everyone had almost come to believe that the sounds of wind and wave would never be changed, the faint rumble of engines could be heard.

This was the moment when those gathered there would stop talking and even moving. Signal lights winked out to sea and back again. Guns at the harbour entrance were manned in case the enemy tried to enter. Slowly the naval ships made for their berths. Voices distorted by loudhailers were taking over from the winking lamps. Mooring lines snaked out, were caught and secured, then engines cut.

On several occasions we joined the watchers and waited with them. Such scenes had to be witnessed. I never imagined that one day I would write about them. Perhaps it was as well, for they should not be forgotten. War was not like a clean and colourful television picture. Lightly wounded men tended to thrash about and moan. The more seriously wounded tried to lie still as movement meant pain. The badly damaged were given drugs which took over from the shock. Those perfectly still and silent had paid the ultimate price for what was termed a routine night patrol.

These routines took place most nights at Dover and elsewhere round South-Eastern England during those terrible times. For those of us on RAF boats, our turn began at dawn. Air/sea rescue work — particularly in support of large scale operations such as the thousand bomber raids to come — could be complex, but each action was to an underlying pattern. Our aircraft would take off and cross a stretch of sea to engage the enemy. Rescue launches were positioned to cover the outward journey in case of engine failures or fighter retaliation. However the home run was much more important as damaged aircraft, often with wounded aircrew, would come down in the sea. The return route was usually different to the outward one, so either the rescue boats had to be shifted or further craft sited elsewhere. In all cases, launches took up position before our aircraft left the English coast and did not leave until their safe return. Thus time at sea could stretch from hours to days and nights when searching for survivors.

The area, which each air/sea rescue boats initially had to reach and remain at, was called the rendezvous position. Once there, engines were switched off and the launch allowed to drift or to use a sea anchor. Both methods altered the movement of the vessel for the worse. Unanchored boats rolled alarmingly and sickeningly, while those using sea anchors tugged and jerked. Being on rendezvous was the condition which tested the morale as well as the stomachs of air/sea rescue crews. Men who were able to take high speed bouncing over the waves, became quiet and turned green after a few minutes of rolling — let alone the long hours ahead. Seasickness is not a pleasant malady in spacious, well-appointed ships. In cramped surroundings, when men are already wet, cold and tired, it can try the best integrated crews. Timber therefore warned Doc and me, as new crew members, in the following forthright manner.

"If you have to bring up your ring, do it in a bucket, down the toilet or over the side. Understand?" We nodded full of apprehension.

The flare-up of operations off Dover required relays of air/sea rescue launches below the RAF fighter aircraft involved. The first incident in which I took part put our rendezvous position about five miles off South Foreland, neither a safe nor a calm area. German planes could pounce at any time but our main problem was having to lie rolling in those disturbed waters where the English Channel meets the North Sea. At least the day was clear and the sun warm so we could stay on deck.

While on rendezvous, crew members remained at or close to key positions about the boat. For example, Skip would be by the bridge and Timber liked to linger near the wheelhouse. Either Jock or I sat beside the engine room hatch in order to start up within seconds. Sparks usually had an extension lead to his earphones, by which he could continue listening for signals without remaining in the wireless cabin. Medical kit was laid out in the sick bay because we might have to deal with every type of casualty from exposure and shock to shrapnel wounds and broken limbs. Such preparations became routine and we put them into effect without thinking. However on that particular rendezvous, we were to witness what was forever afterwards referred to as 'Doc's party piece.'

I did not see Act I because I happened to be in the wrong place at the wrong time. Jock had just relieved me from standing by the engine room hatch and I went forward to relieve myself. The toilet in the sickbay was occupied, so I went further forward to the other in the bows. While I was there, according to the tale told and re-told, Doc swayingly entered the wheelhouse and approached Timber. Noting how Doc's face had become pale and waxy, our first coxswain wisely backed away. Unfortunately one cannot back far in an air/sea rescue launch and Timber soon found himself against a bulkhead. At that moment Doc heaved and a few drops flew with unerring accuracy on to Timber's white seasocks.

"Get away from me," Timber roared, "you awful thing you! Get out of my wheelhouse. You've got a toilet in your sick bay. Go and spew there."

Our enraged coxswain then grabbed a rag, rubbed his sea socks as clean as possible, after which he threw the rag over the side.

That was the end of Act I, which I did not see, though I was now to have a front seat for Act II. On my return from up forward, I sat on one of the sick bay bunks. No sooner had I done so when there were some slithering sounds as Doc fell rather than came normally down the steps. I could see he was not at his best and, when I spoke to him, he did not reply. Instead Doc made straight for the door of the sick bay toilet. He did so in a series of swaying motions to try to counteract the rolling of the boat. There was a vertical handrail beside the door which Doc gripped with his right hand while clawing at the handle with his left. I called to him, saying the toilet seemed to be occupied, but again he took no notice. Doc then began to beat on the locked door and make moaning sounds.

I was about to lead him forward when we both heard the inside bolt being withdrawn. Instantly Doc tugged at the door and wrenched it open. This gave me a view of our seated skipper, trousers round his ankles. A split second later, as the boat rolled, Doc leaned forward and was sick all over the entrapped victim.

As if that was not bad enough, there followed some extraordinary contortions with Doc endeavouring to enter the narrow toilet and our skipper frantically trying to leave it. Somehow they managed. The door shut, cutting Doc off from my view, and our skipper stood in front of me, his condition being well nigh indescribable. One often hears that breeding tells. Well I can confirm this. Skip looked down at himself, glanced at me and said,

"Would you be a good chap and fetch a bucket of fresh water? I seem to be in a bit of a mess."

In the end the skipper needed several buckets. During the stripping and cleaning operations, Doc emerged. Our medical orderly began to apologise profusely for what he had done, but Skip merely asked if he was feeling better and suggested that Doc get his head down. In the end we managed to clean up everything. Afterwards, the skipper and I left Doc lying under a blanket on one of his own sick beds and returned to our stations.

Throughout that day we continued to maintain position off South Foreland. Apart from Doc, who was still supine in the sick bay, the rest of us remained on deck to take advantage of the warm sunshine. As morning turned to afternoon, our appetites began to sharpen and Timber asked Paddy what he could provide.

"Our meat rations came aboard this morning," Paddy replied. "How about all hands helping me to make a real Irish stew?"

The idea appealed to us and we set to with a will. Pots and pans were brought on deck, as were the meat rations, potatoes, carrots, onions and other ingredients. While Paddy carefully cut and diced the meat, the rest of us peeled spuds, scraped carrots and sliced onions. Even Sparks with his long lead and earphones joined in the preparations. Eventually all was ready and put into a large pot with water and some salt. We appreciated that our stew would take time to cook but making it had been an enjoyable collective effort, a good moment to remember in a bad war.

We were generally clearing up, throwing the various unwanted bits and pieces over the side, when movement was heard from the sick bay. It was Doc looking like a ghost and about to perform Act III. He could hardly climb the steps and had difficulty in negotiating the hatch. Then, no sooner was Doc on deck, when he sank to his knees in front of our stew pot. Mesmerised we watched him place his hands on the sides, then he added to its contents.

The strange part was how no one moved or said anything. We stood watching Doc in stunned silence until he had finished. Later it was unanimously agreed that, while Act III had been short, it provided a real climax. Skip kicked the stew pot over the side before asking Paddy,

"Any chance of some bread and cheese?"

In all fairness to Doc I must say that, having passed this stage, he was never again troubled by seasickness. For my part, I would become immune if we were constantly at sea. However, after a rest period or even a short leave, I had to go through the whole acclimatisation process.

The sea off Dover could be vicious as I well remember one day when we were duty boat. A strong south-wester was blowing up the Channel and great waves hammered on the harbour walls. We hoped and prayed there would not be a crash call, so one came. The navy said it was too rough to go out, so we went. We never found the pilot and I do not believe anyone could have lived long in those waters. The waves appeared to be moving in all directions as if fighting each other. Again and again we were flung to one side and we began to fear for our own safety. On returning to Dover Harbour, it was pure luck plus our powerful engines that took us safely through the entrance.

Up to that stage, neither Doc nor I had participated in the supreme experience enjoyed by the rest of the crew — that of rescuing an airman down in the drink. The opportunity arrived at last. The weather was better, the wind had dropped and the sea dead calm. We were on rendezvous well towards the coast of France when two planes, far overhead, began to circle then fire at each other. Suddenly there was a flash, followed by an aircraft falling apart. Nearby, a parachute had opened.

"Do we go?" Timber asked Skip who hesitated. The reason for both the query and hesitation was the area in which the pilot would descend. It lay within that big bite out of the French coastline between Calais and Boulogne. The enemy guarded it zealously and would no doubt be sending out rescue boats of their own. Yet the skipper did not hesitate for long. He nodded to Timber and we went in — to quote an expression of the period — like Errol Flynn. As we approached the spot where we had watched the parachutist descend, it appeared quite empty. A moment later Timber pointed and, looking in the same direction, we saw a black clad figure atop a black inflatable dinghy.

"Looks like a Jerry," the skipper said and he was right. The Luftwaffe pilot had that clean cut and slightly wild look one can still see in the photographs of those days. By now, due to Fancy's delicate work at the wheel, we were practically alongside him. Taff and Paddy released the crash net nearest to the dinghy, then went halfway down to help the pilot aboard. But the German was not having any such assistance. Waving them aside, he caught hold of the net and swiftly scrambled up it. On reaching the deck he snapped to attention, clicked his heels, raised his right arm and shouted,

"Heil Hitler!"

I was at the engine room hatch at the time, on the other side of the boat, so did not see all that followed. What I did see was the pilot flying backwards into the water, then I noticed Timber rubbing his knuckles.

"I say," said the skipper, "steady on."

Taff told me afterwards, they nearly lost the German who was floundering in the water. Luckily Paddy was able to grab him. After that they had to haul him up the crash net and on to the deck as the pilot was still dazed. Judging from the coughing, he had also swallowed a great deal of sea water. While the pilot sat there dripping, I reflected on my first air/sea rescue. It had taken a year of training and countless hours of discomfort to achieve one pick-up, yet nothing seemed to work out as I expected. A long time afterwards I was to ask Timber why he hit the German so hard. Instantly he replied,

"I wasn't going to have him throwing his weight around my boat."

60ft air/sea rescue pinnace powered by three Perkins engines in operational service during World War II. *(Courtesy Group Captain L.R. Flower MBE.MM.)*

Chapter 3

LOWESTOFT NIGHTS

Shortly after I joined the crew, Skip informed me that they were not actually based at Dover, but only attached there for an unspecified period. Their base was RAF Calshot at the western entrance to the Solent and, unless a change of plan took place, they would be returning in due course. He said I could apply to remain at Dover and, if I decided to do so, he would help me. On the other hand I seemed to fit in well with his crew and he hoped I would stay aboard. I opted for staying provided I was wanted and could be of service. It was one of the best decisions I ever made.

No sooner had I decided, than there came a change of plan. The skipper returned from one of his briefings with the news that we would be attached to Lowestoft. We were to go there directly from Dover and act as a relief boat. The powers that be also informed us we should expect attachments to other air/sea rescue bases round East Anglia and further up the North Sea coast.

I was inwardly excited at what this might mean, but kept my views to myself while listening to those of Skip and the rest of the crew. The skipper said he did not like attachments as these usually entailed the new boats doing the dirty work. Timber hinted darkly about Lowestoft being 'a hard base' with Skip and Fancy concurring. Doc was upset because his wife had settled in nicely at Dover. Job, digs and conjugals were all organised and now this had to happen. I was however most intrigued at what Sparks had to say. He belonged to that confraternity — one might even call it a secret society — of wireless operators who continually spoke to each other across the length and breadth of the country. Regarding our new attachment, Sparks commented,

"Bomber Command's taking over. Life's going to get hectic."

He was to be proved right. Meanwhile our immediate aim was to prepare for the journey. Doc obtained permission to go and confer with his wife, leaving the rest of us working non-stop. Rations, stores, ammunition and other items had to be brought aboard. We crossed the harbour in order to fill our fuel tanks. Then the skipper and Timber pulled out charts of the Kent coast, the Thames Estuary and the Essex-Suffolk approaches. Calculations were made, tidal tables consulted and fine pencil lines drawn round or between innumerable sandbanks. The charts appeared to be littered with them. There were also ominous grey areas laconically labelled minefields.

It was Sparks who mentioned that we were not going direct to Lowestoft, but would be spending a night at Felixstowe. He referred to this over a meal and Timber gruffly said it was the skipper's decision — a personal matter. We left Dover without naval niceties. Taff and Paddy quickly cast off and we were out of the harbour within seconds. A choppy sea became calmer once we had rounded South Foreland. From afar the Kent coast looked peaceful enough though we knew it bristled with gun emplacements, tank traps, radar stations and fighter airfields. I completed a stint in the engine room, went to the wheelhouse and found Timber in an informative mood.

"There's Deal," he nodded to port, "and the other way lie the Goodwin Sands. People have walked on them at low tide, also been caught when the tide returned. Holiday makers, picnic parties and the like." Timber said this contemptuously and I nodded vigorously. "That's Ramsgate ahead," he went on. "Lovely base, unlike Lowestoft. Manston's up there — the Battle of Britain airfield."

As our journey continued, I was interested how the water changed around us. Up until then I had thought of the sea as a blue-green liquid. I had not expected colours to alter, not only on account of wind and weather, but because of different coastal conditions. Timber told me that experienced sailors could recognise the water anywhere around the British Isles. If taken there blindfold, he claimed, they could guess the location within a mile or so. I had come to like the English Channel despite its sudden changes. It was more green than blue and even its storms were bracing. The Thames Estuary was simply liquid mud, much diluted, but mud all the same. Then came the North Sea which was different again. I never grew fond of the German Ocean as our enemy called it. Like the Germans it was metallic and menacing. To my surprise, because he had made his living from it, Timber agreed the North Sea could be a terrible place.

The Felixstowe of those days was grim. I remember we cut our speed to nothing while awaiting permission to enter harbour. As the boat rolled, it was a sobering thought that heavy guns — no doubt trained in our direction — could blow us out of the water at any moment. When we began a slow approach, the defence details were revealed. We saw low emplacements with sinister gun slits and the sandbagged sites of anti-aircraft batteries. Closer still and we came upon hundreds of naval fighting ships, Felixstowe being a large coastal forces base. The sailors there looked more like pirates than the jolly tars of propaganda posters. As Fancy aptly put it,

"I'm glad they're on our side."

From Felixstowe, even more than from Dover, these naval forces regularly ranged the Dutch and German coasts attacking just about everything they met. They also landed commandos and raided shore installations. Because of vulnerability to aircraft by daylight, they became hunters at night. Yet the noise of their engines and light from phosphorescent wakes made them adopt special tactics. They would creep into position, lie in wait, then fire torpedoes and guns before breaking off the engagement at speeds of over 40 mph. Such craft however suffered from two major defects. Like our rescue launches they were made of wood and ran on high octane petrol. A single shot could transform them into raging infernos.

On our arrival at Felixstowe we moored beside the other RAF boats based there. After reporting to base and chatting to the crews, it was decided that those of us not on watch could go ashore to sample the delights, if any, of the town. Timber warned us not to tangle with the navy, which possibility was far from our thoughts. We were more interested in the skipper's personal business. Resplendent in his best uniform, Skip emerged from the cabin, gave us a brief smile and vanished up the steps.

Presently Timber, Paddy, Taff and I trudged into town. We were discussing what to do — Timber and I were for a sitdown meal while Paddy and Taff wanted to stand at any bar — when we noticed our skipper waiting by a street corner. He had not seen us, being more interested in consulting his wristwatch, which he did every few seconds. Not wishing to reach him, we remained where we were and waited upon events.

Neither he nor we had long to wait. Along the road came one of those MG sportscars with cutdown sides driven by a blonde goddess in the uniform of a Wren officer. She waved on seeing the skipper and he waved back. The next moment she had stopped, opened her door and treated Felixstowe to a flash of long, black silk-clad legs. Another moment and the goddess was in Skip's arms, pressed hard against him. When they drove away, still oblivious of us, Taff exclaimed,

"Did you see that?"

"We saw nothing," said Timber.

Our night in Felixstowe was a routine wartime one. Timber and I enjoyed a fish-and-chips supper. Paddy and Taff managed to find some sailors who claimed no one in the air force could drink. They were brought back to our boat around dawn by a naval patrol, still shouting and singing. While we were endeavouring to deal with them, Skip sloped aboard. Before retiring to his cabin, he said,

"Let's have a bit of hush." Coming from him, that was a strong reprimand.

Mid-morning found us making our way from Felixstowe to Lowestoft. The Suffolk coastline was flat, uninteresting and seemingly endless. Nothing, not a cliff or a hillock, broke the monotony. All we could distinguish was a continous grey line ruled as it were between the slightly lighter greys of sea and sky. Occasionally a few dots would indicate places like Orford, Aldeburgh and Southwold, but they were almost lost in wide open spaces between them. From the sea, such spaces appeared to invite German landings. In reality, both we and the enemy knew it was not quite so simple. The deceptively deserted Suffolk shores were filled with guns and minefields, supplementing the natural hazards of shifting sandbanks and low lying marshlands. Backing these coastal defences lay the largest concentration of air power ever assembled in the British Isles. There were some eighty airfields around East Anglia with many more further inland. Our little boat was to make a contribution, albeit a life saving one, to this vast machine of destruction.

On that journey however we were more concerned with the increasing hostility of the elements. The clouds sank lower and lower as if about to press us into the sea. Rain, sleet and hail lashed us causing the skipper to crouch in his open bridge. Between showers I put my head out of the engine room hatch and saw that the deck was glistening with hailstones. The icy covering made movement between the wheelhouse and the engine room doubly hazardous. No wonder Timber told us to wedge ourselves in corners and stay put. Falling overboard would help no one.

The sea had become our worst enemy. Winds from the north drove straight at us. The waves grew higher and more vicious. Timber kept decreasing speed so as not to batter the boat and us to bits. Later, on crash calls, we would find out what that was like. Now we were learning about the North Sea and its moods, most of which were unpleasant. As well as driving the waves, the gale force winds whipped spray off their crests to create a curious flying sheet of foam above the surface. This gave the impression that we were travelling at several times our actual speed. The overall effect was weird, even eerie.

At last we drew near to Lowestoft. A sea front, along which stood pre-war holiday hotels, led to the town and harbour. The harbour entrance between two high piers looked far too narrow, apart from being seemingly blocked by raging water. Timber however aimed our bows upwind and the gale brought us back to the vital gap. It was an act of expertise which always fascinated me, though I would not like to have attempted it myself. Within the harbour lay yet another flotilla of naval fighting ships, also the largest collection of air/sea rescue launches I had seen so far in one place. As we moored alongside one of our boats, a crew member greeted us with what sounded more like heartfelt sympathy.

"Welcome," he said, "to Lowestoft."

We soon learned that our launch was but one of several due there to provide support for base crews and to help with forthcoming operations. Rescue activities were to be stepped up considerably because of increased bombing raids on Germany. This we expected, though most of us were inwardly alarmed at the rendezvous positions indicated during our first briefing. Many were 200 miles across the North Sea and much too close to the enemy coast for comfort. Lowestoft regulars visiting our boat told of long hours needed to reach such positions and of having to lie there while the bombers droned overhead on outward and homeward journeys. More often than not, rendezvous waits were followed by prolonged searches extending into daylight hours within areas patrolled by German aircraft and E-boats.

Timber kept nodding his head as if confirming what he had meant about Lowestoft being a hard base.

Largely unknown to us at the time, the early war years had seen the British reeling back from the German onslaught. Saving what was left of our army at Dunkirk and beating the Luftwaffe over England bought time for preparation. Nevertheless the Germans still seemed all-powerful and unbeatable. They had demonstrated an apparent ability to wage war on several fronts, including Russia and North Africa. Their navy, especially the U-boats, were decimating our shipping in the Atlantic. Yet the slow-to-anger British, whom Hitler had instinctively been reluctant to rouse, were stepping up aerial attacks on all parts of the Third Reich.

How had the British turned the tables? This is an aspect of the war about which not enough has been written. The masses of aircraft needed to attack Germany did not just appear. They required a tremendous effort from the civilian population in the middle of a long and what often seemed like a completely hopeless war. As aircraft grew bigger and more effective, they also became much more complicated. The two-engined medium bombers, as flown at the beginning of the war, took around 35,000 man hours to build. The four-engined heavy bombers which succeeded them, needed up to 70,000 man hours, also about as many separate parts. The warplanes were not produced solely in large factories. Production lines were fed with equipment and components made in small workshops including garages and backyard sheds. Most war workers were on a six-day week and one in four on permanent night shift. After work, they went on to home guard, civil defence and other duties. It was an incredible collective response and it worked. By the middle of the war, nearly two million people were engaged on aircraft production. The output of bombers, for example, rose from two or three a month to over two hundred machines in the same period.

The names of the planes they made have receded into the past and are dimly remembered. Occasionally we see one at an air show. They seem slow and angular by modern standards. In their time however, those warplanes represented the peak of aviation technology. As mentioned, bombers began with two-engined Hampdens, Whitleys and Wellingtons, then graduated to four-engined Stirlings, Halifaxes and Lancasters.

The Stirlings, built by Shorts of Belfast, entered service in 1941. This bomber was noted for its immense size and weight. It stood more than 20 ft off the ground, making maintenance an occupational hazard, and weighed over 30 tons. The Stirling was slow with an official ceiling of 16,000 ft which was more like 12,000 ft. It meant that bombs dropped from higher flying aircraft often rained down past the Stirlings and added to their unpopularity.

The Handley Page Halifax proved more successful and popular. Halifaxes had a top speed of 260 mph and a ceiling of 21,000 ft. By the end of the war, there were understandably few Stirlings left, but Bomber Command still possessed some 350 Halifaxes. Air crew welcomed the Halifax at first, yet it was soon to be eclipsed by the RAF's best wartime bomber — the Avro Lancaster.

The highly successful Lancaster was fitted with four Rolls-Royce Merlin engines, each developing 1,250 horsepower. It had a top speed of 266 mph, a ceiling of 22,000 ft and could carry a theoretical bombload of 13,000 lb. I say theoretical because this was progressively increased until Lancasters regularly carried 22,000 lb or 10 ton bombs, known as Grand Slams. The Lancaster was popular with air and ground crews alike and, by the end of the war outnumbered all other types in RAF service. Yet it had an inauspicious beginning. The first Lancaster raid on the 17th April 1942 took place against a diesel engine factory at Augsburg. The raid was a daylight one and seven brand new Lancasters out of twelve were destroyed. Nevertheless they went on to become enormously successful as night bombers.

The aircraft production programmes I have outlined, especially the building of the heavy bomber squadrons, necessitated a network of airfields throughout South-East England. The construction of these airfields is another interesting wartime story. Armies of engineers, labourers, electricians, plumbers and all kinds of specialists were needed for the mammoth task. The standard airfield layout was to have three runways, each about 2,000 yards long and 50 yards wide. These were laid as an equilateral triangle and linked at the corners by a circular perimeter track as wide again as the runways. The amounts of concrete required were colossal, yet this only catered for the basic construction. In addition to runways and perimeter tracks, operational airfields needed huge hangars, overhaul workshops, spares stores, also storage for such dangerous items as high octane petrol and high explosive bombs. After that came control towers, radar stations, operating theatres, dental surgeries and in fact all the facilities of a fair-sized town. The buildings had to have lighting, heating, plumbing, telephones and a host of other services. It has since been estimated that the wartime airfield building programme alone cost the country around £200 million — or several billion pounds in today's money. Putting it another way, the bills exceeded one-sixth of the national budget. The Nazis had built a war machine which they used on Warsaw, Rotterdam, Plymouth, Coventry and continually on London. Now it was their turn to be on the receiving end.

After we had arrived at Lowestoft, it did not take us long to sum up the situation at that air/sea rescue base. I believe Timber put it best when he said,

"The lads here have had a tough time. It is up to us to take the weight off them. Give them a much earned break. Therefore we shall be working to the limits."

At two periods during the war, we worked to the limits. The first was that time at Lowestoft and the second during the invasion of Europe. Yet the two episodes were entirely different, being concerned with night and day operations respectively.

At Lowestoft we would try to sleep during the non-operational daylight hours but, even with the portholes covered, it was no good. Invariably someone in hobnailed boots came aboard and something vital had to be done. So we dozed instead until the sun sank, when all the boats packed in the harbour began to come alive. The navy as usual were going out looking for fights. The air/sea rescue crews pottered about while our skippers went for their instructions.

During the wait for our skipper, Timber would look to the charts, whistling tunelessly as he did so. Up on deck, Fancy, Taff and Paddy moved around restlessly, adjusting a rope here and a gun there. Sparks would be fiddling with his wireless equipment and Doc attending to the medical kit. Down in the engine room with Jock, I checked and re-checked each system — fuel, oil and water; electrical circuits, battery connections, propeller couplings, shaft glands and bilge pumps. We used to go over the three main and one auxiliary engine by the book. Defects could cost lives or commit us to spending the rest of the war in a German prison camp.

Generally the navy received their orders first. While we were waiting and checking, the gun boats began running up their engines. These started with sharp bangs, then settled to deep, rumbling sounds. Their officers would jump aboard — bulky in heavy pullovers, flak jackets, dufflecoats and oilskins. By then the whole harbour reeked from exhaust fumes, unburnt petrol and oil slicks. Suddenly the still pattern of hulls, masts, aerials and gun barrels would begin shifting. These movements were all horizontal, as if everything ran on rails, until bow waves crossing and re-crossing the harbour caused the craft to bob up and down. Collisions seemed inevitable but every boat deftly slipped through the entrance to join the hunting packs.

On our first Lowestoft trip, Skip came aboard and announced we had pulled the short straw. This meant our position in the line of rescue boats strung across the North Sea would be the furthest from East Anglia or, putting it another way, closest to the enemy-held Dutch coast. It also meant a four-hour journey to take up our position before Bomber Command aircraft passed over us on their outward flight.

The usual air/sea rescue practice was for the engineers to have an hour in the engine room then take an hour off, unless there was some mechanical emergency. An hour on duty might not seem like a long time until the conditions under which we operated are appreciated. To begin with, the engine room was about the size of a modern-day bathroom or kitchen, though less than half the height. There was nowhere one could straighten up except through the deck hatchway. Crammed within the very much restricted space were the three powerful engines — two side by side forward and one centrally aft. Collectively, up to 1,500 horsepower was generated in that small zone. The noise did not permit any conversation so, when Jock and I were down there together, we either used gestures or a notepad. Earplugs were obtainable but useless. At the end of an hour, one's head was ringing. At the end of a journey, it was impossible to hear what other people were saying and this condition sometimes continued for a couple of days. Added to the noise, were the usual engine room smells of petrol, oil and grease. Oily water rolled about the bilges and grease had to be applied at regular intervals to bearings so they ran warm but never hot. After the noise and smells came the heat. Parts of the engines were untouchable — that is if one was not thrown against them. The engineer on duty therefore tried to remain on his small hard seat above the rear central engine. The hatchway was directly over his head and from that position he could watch the rows of dials and gauges. There were straps by which the engineer could secure himself if the sea was rough and the boat began bouncing at high speed. Thus he spent his hour on duty

During his hour off, the engineer would pass along the open deck, clinging to the safety ropes, then try to find a spare corner below in which to wedge himself. The restricted space was further reduced by the motion of the boat making the foc'sle uncomfortable at cruising speed. At full speed, the forward quarters were well nigh dangerous. Crew members were often thrown and badly bruised, with broken limbs for the unlucky. Boats did not stop for their own casualties and anyone hurt had to remain in the tender care of the medical orderly until the end of that particular operation. Fortunately such a mishap never applied to us.

Eventually the rendezvous position was reached. I am sure that these spots appeared significant on some planner's wall map. We used to imagine them saying — 'A fine line of boats tonight, sir'. At sea however, it all looked the same, especially during the hours of darkness. Being wartime, we used no external lights. As soon as the engines were switched off and the boat began to wallow in the waves, its interior was barely illuminated by one or two blue lamps.

At this point, it was usual for someone to brew up a strong pot of tea. There were also tins of self-heating soup available with ship's biscuits. Fancy often declared that the soup was for the strong stomached and the biscuits for the hard toothed. Paddy, I recall, liked making himself bacon sandwiches. Between trips he would scour the town butchers and grocers getting scraps to fry. This he did over the smelly paraffin stove using an equally smelly much-burned pan. He would wedge himself in the galley, grinning with anticipation while shifting the fatty mess about with a fork. None of us, not even Timber, were able to linger long in the vicinity whenever Paddy had a bacon frying session. He would then bring the result, stacked between great hunks of bread, to munch beside us.

Thus we waited huddled together, clutching mugs of tea and making desultory conversation. We had to be on instant alert, so talking was the only relaxation allowed us. Each crew had its own way of passing the time. We went in for full accounts of parents, homes, schools and sex lives. Mine was singularly lacking, so far as the last item was concerned, but Doc made up for this by graphical accounts of his conjugals. Afterwards I found great difficulty in keeping a straight face whenever Mrs Doc came on the scene. We did learn much about Eton from the skipper. According to him, its secret was in having only five or six pupils per teacher. Timber told us about life as a fisherman off the cold coasts of Northumberland and Fancy related behind-the-scenes secrets of the West End stores. He claimed that the stores marked up manufacturers prices by a hundred percent, also prices were further increased at sales time. On the other hand, Taff's life had been grim, Paddy's idle and Sparks mysterious. As for Jock, he just grinned.

Not long after our boat had reached the rendezvous position, we would hear the reason for being there. A low droning sound would steadily increase as hundreds of heavy aircraft began to pass overhead. We seldom saw them, but the noise they made was unforgettable. Despite the dangers that lay ahead for them, they droned steadily onwards and eventually their engine sounds died away.

After that came our second and longer wait. The night hours would crawl by and the North Sea cold creep into us. Layers of heavy clothing and hot cups of tea or soup did not seem to help. Our hands and feet became ice-like and it was all we could do to stop our teeth chattering. Conversation died and we either stared into the darkness or catnapped. Meanwhile the night wind keened in the aerials and the restless waters slapped the hull.

The RAF bombers timed themselves to be over their targets around midnight. This gave them more precious hours of darkness to try to get home. By then the enemy was fully alerted and determined to exact retribution for the attacks. There were running fights back to the coast and beyond. Already all radio silence was broken. The air was full of horrific information. Of engines out of action, fuselages pierced, parts of wings and tails destroyed, of systems ceasing to work. Worst still were the casualties — crew members hit, mutilated, burned or dead. Each pilot's aim was to get his plane across the sea and back to his home base. Failing that, there were a number of emergency airfields. If there was a possibility of him coming down in the sea, the information was passed via air/sea rescue headquarters and bases to those boats nearest to splashdown.

We were alerted on our first night out in Lowestoft. A Halifax was in trouble and looked like ditching some twenty miles from our rendezvous position. I was told to start the engines and, although we had been instructed to remain on station for the time being, Timber kept nudging the boat northwards. An hour later he had to nudge it back again because a seasoned Lowestoft crew had nipped in and made the rescue. It was disappointing, yet we all recognised that the main point was to get those airmen out of the cold sea. So our engines fell silent again, more tea was brewed and we continued on watch during what seemed like a never-ending night.

The night ended reluctantly though there was nothing much more to see by daylight, only cold grey waters. We were near the Dutch coast, but the low lying land and a slight mist made it invisible. We hoped all E-boats had returned to their ports and the Luftwaffe gone back to its airfields. Nevertheless we maintained full alert — the skipper on his bridge, Taff and Paddy in the gun turrets and Timber by the wheel. I sat in the engine room hoping we could start up again as it would warm the place. I jumped when signal bells rang and Timber called for full speed ahead. Sparks had received and passed on the always welcome RTB signal — return to base.

Then followed the usual procedure of an hour on and an hour off. During my second off duty spell, I stood with Timber and Fancy in the wheelhouse as we approached the Suffolk coast. Skip called down from the bridge that he could see Lowestoft dead ahead. This feat of navigation seemed a miracle to me. It was equally good to enter harbour, moor up and cut the noisy engines. While our skipper reported to the air/sea rescue base, the rest of us took it in turns to wash, shave and prepare a light meal. We did not eat much for already people were slipping into their sleeping bags. No one expected a deep sleep or one of adequate length. However we had to get our heads down, even for an hour or two, because come nightfall we would be repeating the operation.

We went through the same routine for the next ten nights and days. During that period, we had three crash calls and found nothing. We were then given twenty-four hours off duty and spent our free time in various ways. Skip had another beautiful girl call on him — a stunning brunette this time. Timber opted for remaining on boat watch so he could clean up and write an extra long letter to his wife. Fancy and Jock went into Great Yarmouth, Paddy and Taff decided to try out the Lowestoft pubs, while Sparks did his usual disappearing act. As for Doc, he was concerned for Mrs Doc. She had dutifully come to Lowestoft but was still without a local job and decent digs. She arrived at the boat to discuss it with him and Timber hauled me off to a seaman's mission hall to play several games of snooker, all of which he won.

"That damned Doc," Timber grumbled as he potted ball after ball. "He'll lose me my stripes."

Soon it was back to operations with a vengeance. For two nights running, we were caught up in naval fights. We were lying quietly at rendezvous, sipping tea and minding our own business, when the navy went hurtling past us. Minutes later, they and the Germans came hurtling back again, firing everything they had at each other — also our way. Night fighting at sea was a curious business. One heard engines and saw what resembled pretty fireworks. Tracer bullets would curve slowly and gracefully at first, then go hissing past. On the second night, what looked like huge fire rose out of the sea. It died away almost as quickly, leaving us peering into the darkness. Someone's petrol tank had been hit and exploded. We searched in that direction but found nothing.

When we got back to Lowestoft, gun boats were still limping homewards and entering harbour. Although my colleagues were turning in, I decided to remain on deck and watch the MGB's. The reception of those fighting ships followed the same pattern as at Dover. Medical and salvage teams stood by. The wounded were rushed into ambulances, then maintenance men swarmed over the boats. They had to work fast because by evening every available craft would be needed for the next gun battle. One naval vessel, moored behind ours, had a tarpaulin thrown over the forward cannon. The gun was surrounded by what looked like an armoured-plated dustbin to protect the gunner. A group of sailors stood by the tarpaulin conversing in whispers. Intrigued, I asked one what had happened.

"It's me mate," he replied, pointing to the gun position. "He's still in there with his head blown off."

I went below to avoid witnessing the extraction of a headless corpse. Everyone else was asleep. I tried to do the same without success.

Twenty four hours later we had our first North Sea pickup. During the night, RAF bombers went for German naval installations. These were heavily defended, many planes hit and air/sea rescue launches directed into the area. To begin with, we were told to look for a Wellington which might ditch at any moment. We spotted it flying very low but airborne all the same. It continued over us and later we learned the Wellington reached base, a tribute to its strong construction.

While Sparks was reporting the Wellington, Taff began to point and shout there was someone in the water to starboard. It seemed everyone aboard could see the person but me. They knew how to look for a tiny dot in the shifting waves. Even when we were a few yards from the figure floating in a bright yellow Mae West, it was difficult to keep one's eye on the exact location.

"I think the place for you," said the skipper, as he went past me to where Taff and Fancy were letting down a crash net, "is in your engine room." I was the youngest member of the crew, a mere beginner in the ways of war.

From the engine room, I could hear scuffles on deck with no one saying much. After about five minutes, Fancy put his head through the hatch and told me to wind up the engines as we would be making for Felixstowe.

Presently Jock took over and I went to the wheelhouse. The skipper and everyone else crowded there seemed lost in thought. Normally one or two people would go below to help create room. I glanced into the sick bay and saw Doc standing over a well wrapped up figure.

Later, after we had got rid of the body at Felixstowe, the story came out. The dead man was a rear gunner. He had not been in the previous night's action. He had baled out of a Stirling over a week ago and dropped into the North Sea with only his Mae West. Sometime during the lonely days and nights which followed, he died a cold, wet death. As his life jacket deflated so he sank deeper into the water. A long time afterwards — three years later in fact — Doc started talking about the rear gunner whose eyes had been removed by seagulls.

Shortly after picking up the dead man, there came another lull in operations. Half the crew went on weekend leave — Timber, Sparks, Paddy and Taff. By then Mrs. Doc had found a place in the local hospital, but off duty shared a room with three nurses. Fancy, Jock and I ho-hoed when Doc told us about it and we heard Skip chuckle in his nearby cabin.

"Fortunately," Doc went on, "the hospital is huge. There are lots of rooms empty at certain times — such as X-ray departments and dental surgeries. I bet," he gave a sly smile, "none of you have ever had it in a dental chair."

"I am beginning to think," called out the skipper, "that this boat is no place for a parson's son."

The lull in operations ended when Bomber Command further stepped up its attacks on Germany. The trend was towards raids using over a thousand aircraft with all that implies by way of damaged planes, crews lost and searches for them. As these operations increased, the weather worsened. We were out in gales lasting for days. We were battered by heavy seas which the navy said were downright dangerous if not impossible. We took part in several successful rescues.

One rescue remains in my mind because it was so different, the unexpected happening again and again. It began with a Lancaster making a perfect splashdown — that is if bringing a badly damaged, four-engined landplane down in the North Sea could be called perfect. None of the seven-man crew were hurt and they managed to get clear in their large, inflatable rubber dinghy before the bomber filled with water and sank. We were given their position and, though it was too near the Zuider Zee for our liking, we began a search which we hoped to complete before daybreak.

Dawn found us still searching the area. There was a thick sea mist and the temperature began to drop alarmingly. Within half an hour, the mist had changed to freezing fog, making an already fruitless search all the more difficult. While we watched, our entire launch — deck, superstructure, masts and aerials — became coated with hoar frost. As for the sea, it flattened to a dead calm, with the surface looking like thick oil.

The skipper and Timber discussed what best to do. They decided to switch off our engines and wait for the fog to clear. As there was no wind or a sea running, it was thought our boat would maintain position relative to the dinghy, however long the fog persisted. Timber told of similar situations with fishing vessels off the Northumberland coast. He started to talk about a certain incident, when Taff came rushing into the wheelhouse and reported the sound of voices. Cautiously we followed him on to the icy deck, listened and could also hear men talking. The RAF aircrew in their dinghy were quite close to us. We called and they answered. A few minutes later, seven chilled but happy survivors were aboard, none the worse for their experiences. It was, the skipper contended, a most suitable occasion for opening our emergency rum rations and everyone agreed whole heartedly.

We were enjoying liberal tots and becoming a little tipsy into the bargain, when Taff came to report more voices. Only this time, he added, they were German. The skipper and Timber looked at each other, then shot up on deck. They too heard German voices — barks and grunts Timber called them. Then we heard diesel engines from another direction and knew we were surrounded by E-boats. Our only hope was to remain absolutely silent.

An extremely long hour followed as we lay listening to those voices and engines around us. When at last the Germans seemed to have left the vicinity, Skip thought we ought to make a run for it. Our engines sounded over-loud when I started them. After that, we ran at quarter throttle for about fifteen minutes.

"There," the skipper is reputed to have said, "we've made it." A moment later tracer shells hissed overhead. The enemy was following the sounds of our engines. More tracer went tearing past us.

Apparently Timber acted while everyone else looked at each other. He swung the boat away from the firing, covered half a mile at top speed, then cut our engines. So did the Germans. Half an hour later, we tried moving slowly away and those hunting us did the same. We decided to wait longer, forever if necessary. It was not a comfortable wait because an E-boat could loom out of the fog at any time. One aircrew said,

"They must have reserved our rooms at Stalag."

Morning turned to afternoon. After total silence for over three hours, Skip decided on a trial run followed by an abrupt halt. We re-started our engines and travelled for five minutes before stopping. Again silence. So we set off on a course for Lowestoft, moving at half speed through the murk. Suddenly, like a cliff rising out of the gloom, we saw a large merchant ship at anchor. The vessel was dead ahead of us and Timber had to slam our launch into reverse to avoid hitting it.

There is a pleasant procedure when two ships meet, as we did, and are able to lie alongside. Each person socialises with his counterpart — captains, coxswains, wireless operators and the rest. A merchant navy chief engineer came to view our microscopic engine room, then took Jock and I to his domain, which was about the size of a cathedral. Afterwards we went to his cabin to enjoy, as he put it, a few drams. The engineer was a Scot like Jock and they soon lapsed into their barbaric tongue. A little later, we were all invited to a fine meal with more to drink. There, the aircrew captain sang our praises.

"These air/sea rescue boys are fantastic! Found us in a fog. Eluded the Germans. Then located this ship. Never seen anything like it." We smiled modestly.

The fog lifted next morning. It did so rapidly, the last few shreds whirling about us before disappearing to reveal further magic. We were close to the Suffolk coast, just south of Lowestoft. The aircrew piled back on our boat, filled with further admiration, and we took them to the harbour. Sparks had sent ahead news of the rescue and a truck from their station waited on the dockside. After many thanks the bomber crew left. We were beginning to feel our true tiredness and declared that nothing could stop us sleeping round the clock. Some of us were already in our bags when the skipper returned from reporting to base. He had news for us. We were attached to another air/sea rescue base further up the North Sea.

Chapter 4

CALSHOT REFIT

In the end we were attached to three other bases further and further up the North Sea coast — to Gorleston, Bridlington and Blyth. The weather continued bad and the work hard. The good news was that we helped to save more lives and Timber was able to visit his wife from Blyth. The bad news was we became quite disorientated, somewhat like stateless persons, particularly the Docs. However I learned some interesting lessons from that period. One was every town had facilities in the form of hot baths. Another was a knock on almost any door would result in some kind woman taking in washing. Unfortunately our moving around tended to confuse the arrangements. The crew was fast becoming discontented at all this chopping and changing. Everyone appreciated there was a war on, but could not see why we kept wandering about 'like a lot of bloody Flying Dutchmen'.

"We'll have to tread warily on this one," said the skipper when the matter was put to him. "You know what the RAF is like. Put forward a logical argument, hint that the crew is becoming browned off, and they'll go beserk. Call it mutiny on the high seas. Flog us round the fleet. We must play this their way. I am willing to write a report stating our boat is in urgent need of a refit. I could say we're springing leaks, there are barnacles on the bottom, the engines are clapped out and the propellers coming loose."

"The propellers are all right," I ventured but Timber said firmly,

"They're coming loose."

I do not know what exactly the skipper wrote, but it worked wonders. We were to return to our home base at RAF Calshot for an engine change and general refit. Timber penned an extra long letter to his wife about it and Doc rushed off to put Mrs. Doc in the picture. We were leaving the North Sea for the South Coast.

The weather improved with every mile we sped southwards. Grey clouds which had been clamped down on our heads for months, lifted to reveal that blue sky still existed. Seemingly eternal grey rollers sank to a smooth sea green. We put into Dover on our way round to Calshot, looked up some old cronies and staggered back to the boat in the early hours. Paddy sang Danny Boy plaintively as he descended the dock ladder, stepped on to the boat, crossed the deck and fell off the other side. Knowing he was a good swimmer and drunkards are never harmed, we let down the crash net and left him singing in the harbour. While going below, it was decided not to mention the matter when Paddy managed to join us. At last he appeared, water pouring

Early 50ft pinnace powered by single Gardiner semi-diesel engine. *(Courtesy Group Captain L.R. Flower MBE.MM.)*

off him. Ignoring the sodden apparition, who incidentally sat on my bunk, we carried on an inconsequential conversation until the Irishman could stand it no longer.

"You English bastards," he said, "might be interested to know that the water is quite warm."

"Water? What water?" we asked, then fell about laughing.

Next morning the Channel was Mediterranean blue. We cruised like millionaire playboys along the coast to Dungeness, Beachy Head, Worthing and Bognor Regis. Again, from the sea everything looked peaceful ashore yet we knew the entire area contained minefields and gun emplacements. Whenever we neared a beach, one could see seemingly endless rolls of barbed wire.

After passing Selsey Bill, it looked as if we were making straight for a solid headland until the Spithead approach opened in front of us. Because I had not been there before, the crew kept pointing out places of interest. Ryde and Cowes to port. Portsmouth, Gosport and Lee to starboard. By then we could see Calshot Spit below the New Forest skyline. Lines of buoys lay from Calshot Castle along Southampton Water. A flashing lamp welcomed us home.

We were some months at Calshot, months in which the invasion build-up took place around us. Although we had approached our home base from the sea, the normal way was overland from Southampton. This was done in a quaint local train known as the Fawley Flier. It used to rattle round the top of the Solent via Totton and Hythe, with little halts like Pooksgreen in between, before indulging in a final gallop to Fawley. The village was then small and picturesque, instead of the present vast oil refinery. From Fawley we walked the last mile or so to RAF Calshot, the New Forest rising on the right and the Solent to the left.

There was an upper camp comprising guard and orderly rooms, cookhouse and canteen, living quarters and sickquarters. This stood on heathland bordered by great old fir trees. The road from Fawley continued past it down the spit to a lower camp about another mile further on. As one neared the end, usually propelled by a brisk sea breeze, two red brick buildings rose beside the Solent. The first was station headquarters and the second accommodation for duty crews with another cookhouse. It was rumoured that the two cookhouses held competitions on how best to ruin good food.

After passing the two red brick buildings, the road widened to a great expanse covered in concrete, bordered by hangars and workshops on each side. Ahead lay the castle and slipway with the pier projecting between them. The pier ended in a crew hut next to a small hand-operated crane. This hut had large glass windows overlooking the boat and seaplane moorings. Upon our arrival at Calshot, we had to moor quite a way along the Solent due to the great number of vessels and planes present. We spent our first night among these many craft then, next morning, headed for the slipway.

The object of the exercise was to pull our boat up the slip so that its hull could be cleaned, inspected, repaired and repainted. To achieve this, meant lowering a wheeled cradle down the slip and under our craft. When the craft fitted snugly on the cradle, the two would be secured together then pulled up the slipway. It was a simple enough exercise, but carried out by the craziest characters I had ever encountered.

Heading them was surely the air force's tiniest warrant officer wearing the largest hat. His voice was also out of all proportion to his minute stature, rising effortlessly from baritone to falsetto. He was accompanied by a band of civilian cronies who could have have made a fortune as extras in Treasure Island, Jamaica Inn or Captain Blood. Their villainous faces were not improved by cropped hair, broken noses, missing teeth and eyepatches. Many wore gold earrings. One I recall went barefoot and had downward curving nails on his big toes which he would slash out at passing people. They all shouted and the slipway resounded in what could be described as a kind of verbal-sexual frenzy.

Somehow we reached the top, then came the turn of Calshot's chief engineering officer to make his preliminary inspection. He moved knowingly underneath the hull with the skipper and me in attendance. This officer obviously knew his way around marine craft and was not the kind of man to try to fool. Ignoring the clinging barnacles and dripping seaweed, he made straight for the propellers.

"I can see no cause for the reported vibration," he informed the skipper after a close study of each propeller, shaft and bearing bracket. Without looking in my direction, Skip murmured,

"It's when we're travelling fast."

"Hm," the engineering officer grunted.

He did however admit we needed new engines, also a scrape-and-paint job. While this was taking place, we would have other duties. Skip and Timber were relegated to orderly officer and orderly sergeant respectively. Fancy, Paddy and Taff had to scrape then repaint the hull, while civilian technicians changed the engines. Sparks was sent on a wireless operator's refresher course and Doc had to work at station sickquarters. Jock found himself transferred temporarily to ferry pool, which meant taking new boats to air/sea rescue bases and bringing old ones back for repair. In my case I was to join the Calshot duty crew responsible for maintaining all RAF marine craft in the vicinity.

This meant my moving to the red brick building next to station headquarters beside the Solent. Upstairs there was a spacious dormitory with some thirty beds in it. Views from the windows were panoramic but the whole place was about as draughty as a windjammer rounding Cape Horn. My bed stood between one occupied by a grizzled veteran, who told me he was captain of a refuelling bowser, and a meek little man who worked in the stores.

The bowser captain used to like sucking his unlit pipe while carrying out refuelling work, which sent the midget warrant officer into fits of rage. Both were stubborn and eventually the group captain commanding RAF Calshot was called upon to solve the problem and clash of personalities. He ordained that the bowser man could continue sucking at the unlit pipe provided a piece of white paper protruded from the bowl. Every morning after that, the warrant officer would ask,

"Where's your piece of white paper?"

As for the meek little storeman, I had only been there a couple of weeks when a pair of policemen — one service and the other civilian — came for him at four in the morning. I woke to see my neighbour packing his kitbag and I said to the SP,

"You're like the Gestapo arresting people at this hour. I thought we were fighting a war to prevent such behaviour."

"Simmer down," he replied. "We have to catch an early train. Your neighbour is wanted up north to explain why he married three women."

My working day as duty engineer began an hour later. I would go to the pier and start the marine tender lying alongside. Next I visited each boat on the lines of buoys to run their engines and generally check them over. This I was expected to do on my own having had both deckhand and engineering training.

Provided their batteries were well charged, petrol engines started fairly easily. The diesels I found more difficult until an old hand taught me a highly illegal trick. The trick was to wrap a piece of rag round a long screwdriver, dip this in paraffin and set light to it. To start a diesel, its intake filter was removed and the torch held beside the opening. On turning the engine over by its starter, each cylinder would suck flame and the diesel was running in seconds. There were two points to watch. The rag had to be wired securely to the screwdriver or it would be pulled into the diesel. The second was never, but never, try this trick on high octane petrol engines.

About the fourth or fifth morning of going through such procedures, I had my first mishap. I tied my craft to a seaplane tender, stepped aboard and started the two diesels. After that I went through the general check before returning to the tender. To my horror it had gone — tugged away from an inadequate knot by the tide. The little warrant officer and his piratical cronies I reckoned would at least keelhaul me for the offence. Then I realised something must be done swiftly before my boat drifted too far and caused an accident. Desperately trying to remember my morse code, I snatched up an aldis lamp and signalled the pier.

"SOS. Lost boat."

Almost immediately a swift succession of dots and dashes came flashing back to me. P-U-T-O-N-Y. It meant nothing so far. O-U-R-J. Was it in code? E-S-U-S. Suddenly the message became clear. It read,

"Put on your Jesus boots and walk."

As this flippant message was being transmitted, another boat left the pier to rescue me and the duty tender. The incident had to be reported and I expected the book thrown at me. The little warrant officer however ranted and raved at everyone else. I should have been accompanied. We were supposed to be in the business of saving lives, yet this had occurred. It must never happen again. A notice went up to the effect that the duty engineer must be ferried by another member of the duty crew.

Thinking back to those days at RAF Calshot, the tone was set as at most stations and bases by its commanding officer. He was an amiable group captain noted for his past work on high speed aircraft aimed at winning the coveted Schneider Trophy long before the war. This the air force did on three consecutive occasions, a feat unequalled in the early days of aviation. The lessons learned by Supermarine in building the Schneider Trophy aircraft helped to develop the Spitfire which, in turn, beat back the Luftwaffe and saved Britain in 1940.

Our commanding officer liked nothing better than to stroll round his station, so picturesquely placed between land and sea, with always something interesting taking place. He was invariably preceded by a white bullterrier, a most amiable dog like its master. Someone told me about this but I had not taken much notice. Thus there I was one fine afternoon leaning in a most unairmanlike manner against the pierhead crane. The view across the Solent was so absorbing I did not heed the bullterrier who sniffed around me before wandering off elsewhere. Suddenly a voice behind me said,

"It's all right, I've got the weight."

Wondering who was there and what he meant, I stopped leaning against the crane, straightened up and turned round. To my amazement, I had been joined on the pier by the commanding officer. He was half crouched with one shoulder under the crane jib as if endeavouring to stop it falling into the sea. As the crane was the most solid of structures, I understood the joke but was more concerned with the implications. To a mere airman, a group captain was godlike. Fortunately this group captain merely suggested that I should in future refrain from trying to support the crane, while I nodded agreement to his every word.

The next incident concerning him was connected with our boat being refitted at the top of the slipway. While civilian workers changed the engines Fancy, Taff and Paddy scraped away at the hull removing every vestige of paint as well as marine accretions. I used to help them whenever possible because we were all anxious to end our on-shore existence. At last three new gleaming white engines replaced our old oil stained ones. After they were connected up, the engine room had to be repainted. This coincided with Jock returning from his ferry trip and we worked together. Skip and Timber joined us whenever possible. We began to feel like a crew again.

By then the bottom of the hull had been thoroughly scraped, repaired and treated with various undercoatings. The worst job was the application of red anti-fouling — a thick and sticky substance. Not only was it a devil to apply, removal from one's skin also proved painful. Paddy began to complain that the English had inflicted the task too many times on him. He would squat under the hull, bad-temperedly slapping on the red substance by means of a thick brush. Every now and then, some of the paint would stubbornly refuse to go where it should and bespatter Paddy instead. In consequence the rest of us kept well away from him.

Not so the friendly bullterrier. It was on a routine round and considered Paddy crouching there another friend to greet. To our astonishment, because he liked dogs, Paddy swore at this one and tried to wave it away with his brush. The bullterrier joined in the new game by winding itself round the Irishman's legs. As the dog turned away to visit other friends, Paddy smacked the brush, laden with red anti-fouling, across the bullterrier's white posterior. At the same moment the group captain rounded the corner to see his pet with a bright red stern next to Paddy holding the only brush in the vicinity.

"Why did you do that to my dog?" asked the group captain with more forebearance than the rest of us would have shown in his place.

"He went for me, sir," Paddy replied.

We were amazed that the foolish action should be followed by such a silly explanation. Its silliness must have got through to the group captain, because he burst out laughing. At last, still chuckling, he asked Paddy,

"You say my dog went for you?"

"Yes sir."

"What," tears of sheer merriment appeared in our commanding officer's eyes, "did he really go for you arse first?"

Before the general laughter had subsided, Fancy offered to clean off the paint. I must say the bullterrier was most patient as we padded thinners around his nether regions.

Come to think of it, RAF Calshot was a doggy place in those days. Several of the permanent staff had dogs as did many civilian workers. Their dogs were mostly the breeds countrymen favour — greyhounds, whippets and lurchers — and well behaved they were too. One would come across a dog sleeping at the back of a storeroom or lying under a stationary truck. Each dog had a favourite spot and all were meticulously clean. I remember a certain whippet which would trot from Calshot Spit on to a nearby marshland to squat over a tidal inlet. At the time I thought no dog could be better trained but I was to be proved wrong. The episode also involved Doc with whom I had kept in touch.

As previously mentioned, Doc worked at station sickquarters. His duties were not onerous so he had plenty of time to find a local job for Mrs Doc — at Netley Hospital on the other side of the Solent. The wide stretch of water did not however separate them. RAF boats visited Netley daily thereby enabling the Docs to keep in touch. Inevitably Doc would spend the night at Netley or Mrs Doc come over to Calshot.

One afternoon I went to see Doc at station sickquarters. They appeared empty, which was not surprising, as the would-be-ill in the air force had to report sick before the crack of dawn. Thus I wandered from the general waiting room into one treatment room after another without seeing anyone. Suddenly Doc called from yet another room,

"In here. I'm with a patient."

"Can I come in?" I enquired from outside the half open door, not wishing to invade the privacy of whoever it was being treated.

"Sure," Doc said. "This patient won't mind one little bit."

So I entered the room to see a massive husky dog sitting on a table. The husky smiled at me, as if sharing the joke with Doc, and thumped its tail. Then the dog held up a paw to continue the treatment.

Doc explained that the husky had been brought back to Calshot by an air/sea rescue crew from Reykjavik, Iceland. Its name was Erik and, following the usual quarantine period, it had been taken on the station strength as a motor boat dog. Erik the husky liked nothing better than mucking about in boats, helping where he could. For example, after our boat engines had been started, someone had to look over the transom stern to make sure cooling water emerged. If it did not — due say to a water pump failure — the engine could overheat and seize, a well nigh capital offence in the marine service. Water-watching was therefore one of Erik's duties. He would stand tensely, gazing over the stern and only barking when the water emerged. Erik had other duties such as being a fine foot warmer during cold nights on boat watch and a highly efficient disposer of food scraps.

Unfortunately Erik suffered from an occupational complaint which Doc was now treating. Salt water sores used to develop between his toes and from time to time needed the application of antiseptic ointment. When this happened, his skipper would order Erik to report sick at the upper camp. Erik used to make the journey by himself, hitching a lift both ways. I am happy to relate that, following his war service career and an honourable discharge, Erik went home with one of the crew and lived to a ripe old age.

While Doc and I were chatting, a senior medical officer came into the treatment room. Usually an officer's uniform did much for a man, but in the case of this newcomer he was still a typical family doctor of the old school. After being introduced to me, the medical officer discussed cases with Doc, referring to them as the sore throat, the sprained ankle and the hernia. Then came a change of subject, causing me to prick up my ears.

"We must also," said the medical officer, "prepare ourselves for Operation Keep Fit. Quite a few, I should imagine, will fall by the way side when we er, reach the second phase and there could well be actual casualties."

"My very thoughts, sir," Doc nodded.

The first intimation of what lay ahead was the arrival of six physical training instructors. These were loose-limbed, clear-eyed types wearing white pullovers, black trousers and running shoes. They were always on the move, either trotting about the place or running on the spot. When not plaguing our lives, they occupied themselves with workouts in the gymnasium. The war and our air/sea rescue work meant nothing to them. Someone called them a 'narcissistic bunch of buggers'.

Their arrival was followed by a notice on all boards to the effect that crews would participate in half an hour of exercise a day. Before breakfast! Thus shivering and grumbling we did our arm waving and body bending. Although we felt better for the work-out, no one would admit it. Our moans increased as the PT instructors began pouncing on us during the day. We would see them from the corner of our eyes bouncing along in our direction, at the same time calling out heartily,

"Come on you lot. Tools and paint brushes down. We shall start with some deep breathing, shall we? Breathe in. Hold it. Breathe out. Now some stretching. . ."

After about ten days, the treatment was stepped up to cross-country running. We would be in the middle of some boat job which we considered vital for our return to operations at sea. Suddenly loudspeakers came on, telling us to change into running shorts and vests, then to assemble by the main gate. There waited two or more PT instructors already limbering up, eager to be off. They took a great delight in telling us of the difficult terrain they had chosen like up and down sand dunes or over large rocks. These inevitably produced falls and strains, which meant visits to station sickquarters. I did not succumb, but became adept at accompanying patients for treatment.

At station sickquarters, Doc acted as an assessor before would-be patients actually saw the medical officer. I was embarrassed how he — as a known member of our crew — used to sort out many of those claiming strain and exhaustion as mere malingerers. One patient, who said his back was bad, threatened to break Doc's back the next moment. For his part, Doc was unrepentant. He asserted that regular physical exercises and especially the cross-country runs were contributing towards better health all round. Yet Doc never joined in the exercises or runs himself.

Worst was to come. Within a month of the PT instructors arriving, the RAF Regiment followed them. If these regimental types had come to guard us, no one would have minded. It soon became clear however that their arrival was intended to augment our physical training. The chief officer turned out to be a big beefy man with a red mottled face. Into this he screwed a monocle which popped out whenever he became annoyed. Therefore, within minutes of reaching Calshot, he was known as Popeye.

Popeye was determined to make his presence felt. We were summoned to a series of pep talks which left us profoundly depressed. Stripped of pseudo-military jargon and cliches, what Popeye told us was as follows. We had to succeed in the coming invasion of Europe or the Nazis would rampage over England's fair and pleasant land. Our women would all be raped and we castrated. To prevent such ultimate indignities, we must learn to fight — especially with cold steel. For that was what the Hun feared most. Popeye himself loved playing with cold steel such as knives and bayonets. He gave us graphical descriptions of what happened when cold steel, as he put it, 'went in'. At one training session, he charged a petrified Fancy who simply froze, goggle-eyed at the oncoming bayonet point.

"You know what you'd be looking at now?" asked Popeye after deflecting the point. "You'd be looking at your own guts gleaming in the grass."

Popeye's process of trying to change us into blood-crazed warriors accelerated. We went for firing practice on a local range. We were shown how to handle bombs, grenades and other lethal devices. We were driven far from Calshot, then told to make our way back on foot. Luckily I was paired with Sparks who believed rules were for fools. On our first outing, he chatted up a farmer's wife who, after giving us tea, had her husband take us several miles on a tractor. On the second occasion, we came across a naval convoy of lorries looking for Beaulieu and Sparks guided them to within half a mile of Calshot. The third and last time we just caught a bus. Like in London, Sparks said.

But the dreaded Popeye had a new terror for us. Rumours spread of a battle course being constructed along the coast. Each morning, regimental types headed in that direction with picks, shovels, ropes and rolls of barbed wire. When bribed with pints of beer, they hinted at fearsome obstacles. Then Popeye began hinting as well. Each and everyone of us, he declared his monocle gleaming evilly, would have to prove ourselves. Hard news came at last. Crews were to go over the battle course led by their skippers. Moreover real bombs and live ammunition would be used.

"If you have not already done so," said Popeye, "I would advise you to make your wills."

So far as our crew was concerned, Popeye irritated rather than scared us. His battle course became another nuisance. It was bad enough trying to be seamen, living in cramped conditions and enduring rough weather. Now our lives were being plagued by PT instructors and regimental madmen. Sensing our morale was at a low ebb, Skip endeavoured to cheer us up by saying that the course could be rather fun. Fun or not, Timber asserted, we would do it properly. Then why not all of us, Fancy asked. Why should Doc, a member of our crew, be exempt? Democracy prevailed. It was unanimously decided that Doc would have to don combat clothes like the rest of us, wear full kit, carry a rifle and bayonet for the forthcoming course. It was a decision which roused our spirits and would particularly concern me.

On the day, we assembled by the camp gate looking like members of Fred Karno's celebrated army. Doc stood apart from us staring sullenly into the distance. Supposedly a man of medicine, there he was with soot daubed over his face and bits of bracken hanging from his steel helmet. In a rash moment he told us that Mrs Doc had been responsible for the make-up and flower arrangement. The rest of us shuffled into some semblance of order before heading for the battle course. Tales of its terrors had reached our ears. How we had to cross rivers by ropes, jump into mudpits and crawl through tunnels filled with barbed wire.

Popeye awaited us as did his merry men, also several officers from local establishments who had come along to lob bombs and fire guns. Anything could happen, said Popeye, then enlarged on the various possibilities. There was no guarantee we would survive the course. Likewise there could be no faltering or turning back. Rifle and machine gun fire would be heavy, especially round our heels. In addition there was a tight time schedule for completing the course. If any of us failed or dawdled, all of us would have to do it again.

"And again!" Popeye shouted causing his monocle to leap outwards.

I recall we were somewhat subdued as Popeye's people, loaded with bombs, guns and live ammunition, went to their pre-determined ambush areas. While we were waiting, Popeye ticked Doc off for holding his rifle and bayonet so sloppily. This tirade was interrupted by a series of whistles from along the battle course. It was time for us to line up and, no sooner had we done so, when Popeye pulled out his service revolver, thrust it upwards and fired. The sharp crack set us off like startled hares, Skip and Timber leading the way. The rest of the crew tried to keep together, but we soon began to string out with Doc at the rear and Popeye firing into the ground behind him. Other firing commenced though, strangely enough, it did not scare us. We knew they intended to miss.

To the young and fit, the course was not that difficult. The rivers were streams and the mud pits dry. Thus we ran, crawled and climbed trying to ignore the noises. The worst hazard was a steep downhill run followed by a sharp drop. We had to take this in two's so I slowed down for Doc to catch up with me. When we reached the edge, Doc hesitated despite Popeye doing a war dance behind him. There was nowhere else to go but forward. I jumped, landing on the soft sandy earth and rolling over as we had been taught. While on my back I was horrified to see Doc throwing down his bayoneted rifle before jumping himself. I had to roll one way to avoid the blade, which hissed into the sand beside me, then go the other way to keep clear of Doc's descending battle boots.

Meanwhile Popeye had run down the side of the drop and was screaming abuse at Doc for abandoning his weapon in the face of the enemy. At the same time, the irate regimental commander was thumbing bullets into his revolver as if about to carry out a field execution. I snatched up Doc's rifle, thrust it into his reluctant hands and we began running again. At least the end of the course was in sight. There stood Skip, Timber and the rest of the crew watching our approach. A single marksman lying on a nearby hillock remained to fire near us. I had just glanced back at Doc, who was dropping behind once more, when something sliced across my right shin. When I paused to look down, Popeye came running up to me brandishing his revolver.

"Why have you stopped?"

"I think I have been hit."

"Nonsense!" he shouted. "If you had been hit, you'd be on the ground, writhing in agony."

He then followed my gaze to the alleged hit. Blood was beginning to seep through the trouser leg, also I could feel it oozing inside my boot. We were both staring when Doc reached us. Without hesitation Doc handed his rifle to Popeye and reverted to being a medical orderly. All firing had stopped by then and our crew, as well as other personnel, converged on us. I was not in any pain and the whole incident seemed to involve someone else. This ridiculous dreamlike state continued while Doc swiftly removed my boots and trousers. How crazy, I kept thinking, which it was.

The wound thus revealed fascinated everyone, especially Doc. There on the front of my right shin was a swelling exactly like the back of a tablespoon. Across the swelling lay a deep groove, as if gouged out by a half round wood chisel, and the blood was flowing freely from that. After some judicious pressing about the afflicted area, Doc declared nothing seemed to be broken but he advised a light tourniquet. He suggested a medical trick of the trade, using my underpants with a stick turned through the right leg hole. Fortunately Skip produced a silk scarf. By then the battle course first-aid squad had abandoned their card game and arrived with a stretcher.

So I became the crew's first casualty — on a Calshot battle course. Everyone visited me at station sickquarters. Skip said there should be a special medal for airmen who tried to be seamen then were wounded on army manoeuvres. Timber told me there had been an enquiry where the accounts officer, who had volunteered to shoot around us, admitted to having forgotten his glasses.

Among my visitors was the accounts officer full of apologies. He kept bringing me cakes his wife had baked, but the only problem was that she could not cook. Even the birds outside my ward window rejected the crumbs. Popeye came as well to assure me that, due to his tightening up the battle course procedure, the mistake could not possibly happen again. Nevertheless, he concluded, we had to toughen ourselves and even become casualties if we are going to take on the Hun.

During the shooting incident, our boat was nearing the end of its refit. The hull and superstructure shone with new paint. The ship's bottom was covered with red anti-fouling and would receive a final coat, immediately prior to going down the slipway, so as to enter the water wet and discourage other marine creatures. The engine room contained three spanking new engines which Jock and I fussed over at every opportunity. By then the civilian workers were beginning to leave us alone and we put the interior to rights after their invasion. Fancy was particularly critical of scratches on the varnished woodwork. Also our cooking utensils had either been burnt black or used as paint pots. We were all determined to return the vessel to a pristine condition.

It is not generally appreciated how many articles are needed to keep a boat operational. Fancy had a book-long list and there was another for me covering the engine room. Because so much had worn out or disappeared, we were invariably at one of Calshot's many stores indenting for ropes, boathooks, light bulbs, paraffin stove fittings, sleeping bag sheets, soap, buckets, sparking plugs, waste rags — our requirements seemed endless. If we had any difficulties with storekeepers, who seemed to believe that shelf items were there not to be used, Timber would throw his weight behind us. He was however still caught up in other camp duties and Sparks went on yet another course in Southampton. Most of the work therefore was left to Fancy, Paddy, Taff, Jock and me.

At the end of the working week, Fancy and I used to go into Southampton. We would do what shopping was required, have a meal, then see a film. Afterwards, there was about an hour before we caught the Fawley Flier and we would fill this by having a quiet drink. The problem was finding somewhere quiet. By that time, Southampton was filling up with Americans who took over most of the city's amenities. Fortunately we discovered a good grade hotel and a comfortable corner of its lounge bar in which to enjoy our drinks.

Fancy and I had not been there many minutes when Sparks entered with an extremely attractive woman. She was, I gauged, slightly older than him, yet still slim and vivacious. The lady wore a saucy little hat perched on top of her blonde curls, a full length fur coat and her shoe heels were so high it looked as if she was walking on knitting needles. Sparks saw us and waved before steering his companion in another direction. For a while we lost sight of them, that is until a group of other people left. By now, Sparks and the woman were seated, engaged in obviously intimate conversation.

I remarked on this to Fancy, but he did not seem interested. Next I noticed that the lady, who had removed her gloves, wore a wedding ring and I asked Fancy if Sparks perhaps was married.

"No he is not," said Fancy sharply, "and don't you ever go following his example. Sparks works on the fact that the world is full of married women. This especially applies in wartime with thousands of husbands away from home. So he takes advantage of women who are, I must add, only too ready and willing. Like that one."

Sparks and his lady friend did not stay long. He waved to us as they left and the woman gave a wide smile. I remarked to Fancy that she looked pleasant enough and had a touch of class. This made Fancy turn sharply and give the person in question his floor walker look.

"No class there," he stated. "Definitely none. For a start, she's wearing cheap shoes."

"But there's a war on. . . I tried to defend her and was firmly told,

"Whatever the circumstances, a real lady always wears expensive shoes."

I was to learn a little more about class a few days later. As I had done at RAF Locking, I brought my bike to Calshot in order to explore the surrounding countryside. I was particularly interested in the New Forest which began just outside the camp. Whenever possible I would be out and about, visiting such places as Beaulieu and Buckler's Hard; Lymington, Brockenhurst and Lyndhurst. Because these explorations occurred in the late autumn of 1943, I always think of the New Forest as full of golden trees distributing coins along the heathland paths. I was following one such path on a Sunday afternoon when I encountered a lady of undoubted class.

She suddenly emerged from a copse on a beautiful chestnut mare which shied then danced away from my bicycle. As I came to a stop, the lady laughingly restrained her mount and at the same time told me not to worry. She formed part of the autumn scene with her golden hair and bright yellow pullover. However, despite her doing full justice to the pullover, the feature that caught my attention was her footwear. While the fine leather riding boots were not new, they fitted perfectly and had been polished to a mirror finish. On seeing them, I reflected on Fancy's dictum about real ladies being distinguished by their footwear. At that moment a man's voice, I did not recognise at first, called,

"Bikes should give way to horses — like steam to sail."

It was the skipper emerging from the same copse. He was mounted on a magnificent grey and also dressed for riding. Skip then made the introductions. The lady was called Jennifer. She lived nearby and they were just returning from a gallop through the forest.

"Returning for tea," said Jennifer. "Would you care to join us?"

I thanked her while declining, but she insisted. Thus I came to cycle behind them through the autumn tinted woods. The late afternoon sun made them and their horses look as golden as the tall trees and fallen leaves.

Presently we came to an open gateway and followed a track, which ran beside a succession of harvested fields, to a stable block behind a big house. There the skipper and Jennifer dismounted, then attended to their horses. I was content to watch, recording the scene in my memory. There was a curious timeless quality about it, as if the war did not exist and none of us were involved.

We entered the big house via a tack room and a series of passages. Eventually we reached a hall with a black-and-white tiled floor. A staircase curved overhead to a balustraded landing. Beside the staircase were large oil paintings in heavy gilt frames. Each portrait was of an important looking person and everyone wore naval uniform. As I studied them, an elderly man with piercing blue eyes came out of a hallside room to greet us. He wore a naval battledress with no insignia. Skip did not mention rank during the introduction, but I later learned the man was a retired rear admiral. Then I was welcomed by the lady of the house in a huge room with floor to ceiling windows along one side. A log fire burned in the fireplace that took up most of the other side. Tea was produced by Jennifer.

For my benefit perhaps, they discussed air/sea rescue work. Jennifer remarked she often plotted our boats and it turned out she was in the Wrens. It also turned out there were more brothers and sisters, all serving in the Royal Navy. One or other usually managed to get home at weekends to exercise the horses and, as Jennifer's mother put it 'to stop the old place from falling down.' I often wondered what became of that family in the New Forest. They were people who had served their country for generations. No doubt various members suffered, were killed or simply lost at sea, yet the line continued until — I would guess — defeated by the taxman.

It was getting dark by the time I managed to escape from their hospitality. I switched on my cycle lights, but the front lamp was of the wartime blacked-out variety and did not help much along the forest paths. At that period the South Coast was supposed to be England's front line, yet I did not see a single person during my journey. At Calshot I met Timber who invariably seemed to be orderly sergeant. He invited me into the guardroom and I shared a fresh brew of tea with him. We talked about the skipper and I told Timber of the people I had met. Then he told me our boat was due to go down the slip during the coming week. We were to live aboard again and there would be sea trials to prepare us for the war which lay ahead.

"It's going to get worse," Timber declared, "before it gets better."

He was to be proved right, but what happened was nowhere near our expectations.

Rescue craft 165 on cradle at top of slipway for cleaning and refitting between operational duties. *(Courtesy R.A.F. Museum, Hendon).*

Chapter 5

FESTIVE SEASON

The re-launching of our boat brought us together again as a crew. Everyone concerned turned up for the occasion. The skipper stood by the pier smiling. Timber was on deck accompanied by Fancy, Taff and Paddy. Sparks put in an appearance, talked about new types of radio sets, then went into a nearby hut which served as local headquarters of the wireless operators secret society. Doc also came to tell us how he would not be able to live aboard until we were fully operational. As for Jock and me, we waited to one side because the new engines could not be started and run until the boat was down the slip and in the water.

There was the usual shouting and swearing. The tiny but histrionic warrant officer did, in our skipper's parlance, "a Richard the Third" . The piratical shore gang appeared to be attacking the new paintwork and Paddy beating them off. Despite the utter confusion, gravity was on our side drawing the boat cradle into the water. Slowly but surely, the level rose over the final coat of anti-fouling which Fancy, Taff and Paddy had been applying since before dawn. At last the hull floated clear of the cradle. Mooring lines held in readiness were used to pull the launch alongside the pier. The next requirement was for us engineers to step aboard and start the engines. Would they or would they not fire?

Jock was worried about this stage of the proceedings. As he said, in what was a flight of eloquence for him, new engines were buggers. I was not so worried. We had fully charged batteries and all three engines fired within seconds of pressing their starter buttons. While they were warming up, we re-checked connections. Presently Fancy put his head through the hatchway and mouthed the obvious question. Were we ready to go to our moorings? I gave him a thumbs-up and within seconds we were purring along the Solent. Soon, I hoped, we would be operational once more.

In order to become operational, we had to work the boat up to a high level of efficiency. This applied to all aspects, including deck gear, wireless and armaments. Our new engines had to be run-in, then undergo speed tests. These were held off the Needles. Other refitted boats joined us on measured runs towards Poole Harbour followed by navigational exercises round the Isle of Wight. The autumn went by and December arrived before the skipper and particularly Timber was satisfied.

Our aim was to be posted to an air/sea rescue base on a permanent rather than a temporary basis. We had experienced too many attachments. Now we

Three rescue target towing launches, Nos 2756, 2772 and 2760 exercising off Plymouth.
(Courtesy R.A.F. Museum, Hendon).

wished to form part of an establishment of about eight boats and crews. Foolishly we believed this was going to happen when eight crews were summoned to the marine craft section. There we were addressed by a jovial squadron leader with handle bar moustache and a press-on regardless manner.

"You chaps," he boomed, "have worked jolly hard getting your boats serviceable. Good show! Next year the balloon'll go up, as we all know, with a bloody big bang. However this year has still to run and there is another quite different task required of you. The Far East boys want eight high speed launches shipped out to them. Don't panic! Neither you nor your refitted boats will be going overseas. Not just yet. There are eight brand new launches in the air/sea rescue pool at Frampton, Gloucestershire. You are to go there by road tomorrow morning, take over those boats and run them to a large merchant ship on the Clyde. The eight launches will be made ready for shipment, then hoisted aboard the freighter. After that is done, you will return here by train — back to your own boats. It will mean travelling up the Irish Sea, whatever the weather, because the merchant ship is waiting to leave. It will also mean being away from here and no chance of home leave over Christmas and the New Year period. But what better place, chaps," the squadron leader stroked his moustache, "to spend the Festive Season than

Glasgow?" A hollow groan greeted this observation.

Moving eight crews and their equipment from Hampshire to Gloucestershire, then to Scotland with eight boats, might appear a difficult exercise. The RAF however was in the moving business. Each crew was given a large packing case into which we put our kit, sleeping bags, tool boxes, ropes, fenders and other gear, topped by our rations. These cases were placed on trucks and early next morning we climbed aboard and sat on them. Soon we were travelling through the frosty countryside to Salisbury, Marlborough and Swindon. We stopped at Swindon for refuelling and hot drinks before moving on to what was once a Roman road running like an arrow over the Wiltshire Downs and Cotswold Hills. This road descended to the abbey town of Cirencester then rose once more, higher and higher, as if we were making for the roof of the world. We were in fact on top of the wolds, another land of butteryellow stone walls and farmhouses, large barns and small cottages. I remarked to Jock, who had made the same journey on a former ferry trip, that it was a fine country and he said,

"Wait till you get to Birdlip."

We stopped at this place called Birdlip, perched on the very edge of the Cotswold escarpment. The morning mist had cleared and the winter sunshine revealed a spectacular scene. The great Severn valley lay far below us as on a map. We could clearly pick out the largest river in the British Isles winding its way from Worcester and Tewkesbury to the Bristol Channel. We could see Gloucester Cathedral reaching like a hand heavenwards and everywhere there were hills. May Hill with its coronet of trees. Bredon Hill of which Housman penned an immortal poem. Cleeve Cloud, Leckhampton, Cranham and Painswick Beacons. The undulating Malvern Hills and the Forest of Dean ranging westwards to the mountains of Wales. One day, I thought, I shall return to and live in this lovely valley, which I subsequently did.

After Birdlip, our small convoy of trucks descended to the valley floor and entered Gloucester. We went straight through the 2000 year old city, past its New Inn built during the 14th century and the magnificent 225 ft high cathedral tower erected in the 15th. Eventually the city lay behind us and we were on the road to Bristol with the River Severn on our right. Also on our right was the 16 mile Gloucester to Sharpness canal. About halfway along, stood the village of Frampton. We reached it as the early December evening closed in. The road ran between wide green swards behind which were some fine looking houses. We were still taking this in when the trucks stopped and another interesting scene attracted our attention. There on a stretch of water in the heart of the Gloucestershire countryside rode a line of air/sea rescue boats. The half light and their grey tarpaulins made them look more like ghost ships.

Excited barking told us that Erik the husky, who was with his crew on another truck, had spotted the boats. A moment later, he was hurtling along the footpath, jumping for joy. While he explored, each crew manhandled its packing case on to the launch allocated. RAF personnel had come from nearby Moreton Valence airfield to help. They brought cooked food and helped us settle aboard. Once we got the heaters going, life became rosier. Life improved still further when we visited the local inns. We tried scrumpy and decided it was dangerous stuff. We played a form of skittles where the ball is suspended from a stick and were roundly beaten by the locals who kept nodding, smiling and saying,

"Oh ar!"

Eventually we returned to our respective boats, slipped into sleeping bags and instantly fell asleep. The next instant it was morning with Timber in a dazzling white pullover shaking us awake. I went on deck as if to confirm that yesterday's scene still existed. Yes, there we were miles from the sea, but preparing to take eight launches from Gloucestershire to Glasgow. Seeing me, Erik leapt across from the next boat. Timber, who joined me on deck, remarked he would have the husky on his crew any day. An hour later we were gliding between fields to Sharpness Lock which let us out into the Severn. At that point, we could clearly see Taff's homeland.

"The Valleys," he informed me, "lie over the other side of those mountains."

"You could take the dinghy and escape," I said jokingly.

"No boyo," Taff replied, "I've escaped already. From the Valleys and the pits." Notwithstanding, Taff treated us to some Welsh singing as we sped past Sully and Barry Island.

That day we also passed the beaches of the Gower Coast and crossed Carmarthen Bay to Milford Haven. It was then a Sunday evening and the town was not exactly swinging. Hymns drifted from the chapels with everything else tightly shut. What concerned us more was the weather. The glass had dropped and the wind was rising. By Monday morning the haven waters were being lashed by gales and we could all imagine what the seas outside would be like. Our captains conferred on the quayside. They did not want to go but their orders were explicit. The eight boats must reach Glasgow as soon as possible. Our skipper informed us of the inevitable decision, adding that we were in for a somewhat rough day.

Even before reaching St Ann's Head, we hit heavy seas. Or rather rollers piling up in St George's Channel hit us. Ours was the second boat and, gazing back from the engine room hatch, I could see the line already dislocated. Hulls pointed this way and that. Bows rose high in the water one moment, then seemed to disappear beneath the waves. After spending weeks in and around the Solent, I felt seasick. I was not the only one. Sparks suddenly leaned out of his wireless cabin and what he produced came whipping back towards me on the wind. Fortunately I ducked in time. After that I recall sinking to the engine room seat and laying my head on the armour plate backrest. It was not the softest of pillows.

When Jock came to relieve me, I went cautiously forward clinging to the safety lines because our craft was corkscrewing in the water. It laboriously climbed each roller and, on reaching the crest, would be caught by the wind. We remained balanced there for a second, followed by a slithering rush into the deep green valley. This Timber told me was the most dangerous stage. Boats with powerful engines had been known to plunge their bows into the next roller and keep going under. As it was, our nose did dig dangerously. Every time, one wondered for how long. As the weight of water came rushing across the foredeck and over the wheelhouse windows, it was like being in a submarine commencing to submerge. Yet, seconds later, we could see out of the windows and the nose was climbing again. At the top of one such ascent, Timber pointed to some storm-lashed rocks and said,

"Bishops and Clerks."

I did not understand him at first but, on referring to the chart, saw that those desolate outcrops, white with foam and bird droppings, were indeed named the Bishops and Clerks.

It was to be a long and uncomfortable day. After rounding St David's Head, we struck out across Cardigan Bay for the north-western tip of Wales. All signs of land disappeared and the driving rain added to our discomfort. No one was able to use the galley, so we made do with tins of self-heating soup — at least those did who could stomach the contents. By then none of the other boats could be seen, though Sparks kept in touch with his wireless cronies. We pressed on through the seething channel between Wales and Ireland. Engine sounds rose and fell as the propellers came out of the water before digging in again. We passed one lighthouse at the tip of the Lleyn peninsula, soon followed by another off Holyhead. Our destination was the Isle of Man and that last stretch in the Irish Sea proved the roughest. We had to cut our speed right down which added to an already overlong day.

When I am asked if I have been to the Isle of Man, I usually reply, "Well yes, but . . " We arrived late on that stormy winter evening, cold and wet. Miraculously the other seven launches also made it. The RAF had laid on a hot meal at a harbourside cafe yet none of us, not even ex-sailors like Timber, could do the food justice. All we wanted to do was slip into our sleeping bags and hopefully not worry overmuch about a similar journey on the morrow.

As it turned out, the next day's trip was not nearly so bad. The wind still blew, the sea remained unsettled and the rain turned to sleet but, on reaching North Channel, the coast of County Down, Northern Ireland, gave us some protection. We saw snow on the Mountains of Mourne, then along the Antrim Hills. By that time we were in sight of Ailsa Craig and about to enter the magnificent Firth of Clyde. Suddenly a tremendous battleship bore down on us and hurtled past escorted by a pair of destroyers. Not long afterwards, with our eight boats again travelling in line astern, we passed between the Scottish Isle of Arran and Ardrossan on the mainland. After Millport, Strathclyde packed with ships, docks and cranes seemed to close around us. Our final run-in should have been past Greenock and Dumbarton to Glasgow, but at this late stage we received new orders. Word came through that the dock, where our boats were to be moored prior to loading on the merchant ship, could not accommodate us until morning. We were told to turn around, go back past Helensburgh and moor at the northern end of Gare Loch.

It was bleak up there. A ring of what the map called Lowland Hills were high and capped with snow. The only objects visible in the empty waste of water were some battleship buoys. These looked like monstrous rusty barrels and were sufficient in number to moor our eight boats separately. We did so and switched off our engines, yet still could not relax. The gale kept on howling, the sea loch remained choppy and every few seconds each boat snatched at its moorings. One had to anticipate the movements involved — a series of lurches followed by a sudden jerk — in order to remain upright. In consequence, as the evening wore on, we became somewhat bad-tempered. The meal we eventually managed to knock together was none too special and a game of cards started afterwards was soon abandoned. By mutual consent, we turned in for the night and tried to fall asleep between jerks.

Sometime later, I heard Fancy asking Paddy if he had moored up properly. Paddy gave an affirmative grunt before turning over in his sleeping bag. Not long afterwards, Fancy repeated the question and Paddy gave more or less the same sort of grunt. All this while, the wind increased and the boat continued snatching. As if to make matters worse, rain started drumming on the deck.

"Paddy," Fancy enquired once more, "are you really certain you did the job properly? It doesn't feel right to me and we don't want to come adrift."

What occurred next is hard to describe. It started with an infuriated roar from Paddy, then he was out of his sleeping bag presenting the fearsome sight of a mad Irishman in longjohns. Paddy snatched up one of our heavy service torches and for a moment I thought he intended to batter Fancy senseless. What he did do, was much more interesting. Using his free hand, Paddy hauled Fancy off the bunk, then out of the bag. In the latter process, the bottom half of Fancy's midnight blue silk pyjamas fell down around the wearer's ankles. Our second coxswain tried to retrieve them while being dragged to the forward ladder and hatch. When Paddy threw open the hatch, an icy blast blew through the boat causing howls of protest.

"What the hell's going on?" Timber shouted but no one answered him. We were more interested in the sight of Fancy's bare white legs vanishing upwards through the hatch.

Timber selected me to render assistance. Hastily we shrugged on oil skins because the weather outside sounded dreadful, then went up to the forward deck. This was like a skid patch, what with ice and more hail falling on to it. I was content to let Timber go first and I clung to every handhold available. Ahead, at the very tip of the bows, were three struggling figures.

I could just make out Timber and Paddy who seemed to be locked in mortal combat. Each had a torch and belaboured the other without either noticing the blows. At the same time, Paddy was trying to concentrate on his main task of pushing Fancy's head over the side. He had largely succeeded because all I could see of our second coxswain were his bare legs and they were having what looked like convulsions. This was due to Paddy holding him by the throat. If Paddy had released Fancy at that stage, then slow strangulation would have been replaced by swift immersion in the icy cold waters of Gare Loch.

"There!" Paddy bawled above the wind and waves. "Now do you believe I moored up properly?" Using the torch he alternately directed its light on the battleship buoy and hit back at Timber.

As Timber wrestled with Paddy, I tried to grasp Fancy's legs and received a kick for my pains. By that time I was soaked to the skin and began to wonder whether my RAF career would be terminated by double pneumonia or being kicked senseless into a Scottish loch. The situation only changed for the better when Paddy decided he had made his point. Fancy was pulled back on deck, picked up and carried below to his bunk. Even during the reverse operation, the Irishman never seemed to notice us.

When we were safely below, the skipper and the rest of the crew listened incredulously to Paddy's version of the incident. He was merely putting Fancy's mind at rest and nothing would budge him from this plain-to-him reasoning. In the end we gave up and from that moment everyone began laughing. We laughed ourselves silly and the generous rum rations doled out to combat any ill effects added to our merriment.

The stormy weather subsided during what was left of the night. By dawn we were cocooned in a thick mist with moisture forming and dripping everywhere. None of the other boats were visible but we could hear their crews talking, also Erik barking. Slowly we crept down Gare Loch, past Helensburgh and somehow located the Glasgow dock awaiting our arrival.

Here we found that the task ahead of us was not going to be a simple one. True the merchant ship was ready and waiting. Its holds were full and hatches battened down to take eight launches on top. Cradles to carry the hulls stood along the dockside and a colossal crane loomed overhead to do the necessary hoisting. Detailed discussions took place between our skippers, the first mate of the merchant ship and an RAF transport officer. There were also miscellaneous dockers and cranemen waiting to work with us.

The task, it was eventually decided, had to be carried out in three stages. First each air/sea rescue launch would be lifted out of the water and put on its cradle. Next we would work through a list, produced by the transport officer, to protect the boats during their sea voyage. Finally, each craft would be lifted aboard and considerable carpentry carried out to ensure they remained in position. The upshot of this was that, even with the crews helping each other and the dockers helping us, we would remain in Glasgow well beyond the dreaded Hogmanay.

During this period we were accommodated at an RAF hotel in a district with the ominous name of Gallowgate. The hotel proved comfortable enough, if a little basic, but the locality was none too salubrious. Men wearing pulldown caps hung around the streets, women in shawls criss-crossed from mean houses to corner shops or pubs, also there were kids, hordes of them, in real rags. It was not a pretty place and particularly depressed Erik. The husky enjoyed the docks during the day but, come evening he lay in the hotel lobby gazing gloomily through the glass door. However Erik had the knack of winning friends and influencing people. The RAF transport officer lived in the country outside Glasgow. He invited Erik home and the husky readily went off in the officer's car to explore the countryside and be spoilt by the family. They wanted to keep him though, in the end, Erik returned with his crew to Calshot and further adventures.

Our Glasgow days fell into a pattern. At some unearthly hour, an RAF service police sergeant would go round the hotel banging on doors. We then had twenty minutes to wash, shave and make our beds; another twenty minutes to eat, after which the trucks were ready to take us to our day's work. It was dirty work in filthy surroundings. At least we could return to our hotel at the end of the day and wallow in hot baths. The dockies who worked beside us had no such amenities. Their living quarters, one could scarcely call them homes, lay right by the docks and shipyards. All one could see were terraces of cramped little houses. I visited several such places and marvelled how families had ever been raised in them. There were four tiny rooms, two down and two up, a cupboard of a kitchen tacked badly on the back and an ash path leading to the outside lavatory. More often than not, two or more families shared the same facility. Gardens, grass and flowers were non-existent. Little shops open all hours served the community, as did sawdust-strewn bars.

We usually went to one of these bars during the lunch hour for hot pies and drams to keep out the chill. It was not long before the dockers accepted us, maybe because we insisted on buying the rounds. They were fascinated by our skipper. He was obviously from a different world and several dockers quizzed us about him. Who was he? Where did he come from? What was he doing here? Jock usually gave them the answers as he seemed to understand their raw dialect. Skip was also influenced by what he saw. He would gaze about the grim scene, rub his chin and be lost in thought. One murky morning, he went away to visit a foreman who was laid up at home with chest trouble. On his return, the skipper really looked depressed and later voiced his thoughts to some of us.

"These people are the salt of the earth. Honest, intelligent and hard-working. Yet what has been their reward? The minimum. It can't be right."

My main recollections are of nights spent in Glasgow. We returned to our hotel around six in the evening for baths and the real meal of the day. Afterwards we went to the city centre, usually in fours or fives, because Sauchiehall Street was no place for solitary beginners. It was nearly Christmas and, despite wartime restrictions, most of the shops, offices and houses did their best with decorations. The main areas of action were the pubs, dance halls and service canteens. These were seething with revellers, mostly Yanks and their girl friends. The Glasgow men naturally resented this and fights were frequent. Vicious fights would break out in seconds with people and places being smashed literally to bits. Worst of all was when two women fought over a man, hissing and scratching like wild cats. One learned a lot during those Glasgow nights.

I remember a certain evening in one bar. Four of us were sitting well back on a wall seat watching the action while trying to keep a low profile. There had already been several noisy Yanks ejected by the snowdrops or military police with white helmets. There had also been a bearded old man prophesying that the end was nigh — until two barmen propelled him outside. Then came the classic conflict of big guy versus the little guy. The big fellow was an American sergeant whose beer belly hung over his belt. He lurched against people on his way to the bar counter where stood the little fellow, a Scottish soldier probably home on leave. Other customers were waiting to be served but the big Yank pushed them aside, leaned over the bar and caught hold of the assistant.

"Gimme," he demanded, "a large whisky."

Those waiting grumbled, yet the soldier only smiled. After the American received his drink, he turned to the soldier. He informed the Glaswegian, in a voice that silenced the rest of the noisy bar, how the Yanks were over here to fight the enemy and screw the women. Again the soldier smiled and, when his turn came to be served, he ordered a bottle of brown ale. This slayed the sergeant who called it brown piss only fit to be drunk by pansies. And so it went on, the soldier smiling and the American baiting him. When the soldier poured half the bottle into his glass and started to drink, the big sergeant thumped him on the back. Still smiling the soldier wiped brown ale and froth from his face and began to refill his glass from the rest of the bottle.

While pouring, the soldier suddenly rammed the base of the bottle against the bridge of his tormentor's nose. Automatically the American raised both hands to his face but he was not in time to avoid the brown ale, with its glass, thrown at him. The position of his hands left an unprotected stomach and the smiling soldier drove a fist into it. Before the Yank could cope with the third attack, a knee came up in his groin. The resultant agony caused the American to bend down which was exactly the right angle for a rabbit punch to the back of the neck. Barely four or five seconds had gone by and there was the big fellow writhing on the floor. As for the small fellow, he was obviously accustomed to such situations. The soldier calmly lifted the bar flap, walked through a door at the back and disappeared. When the snowdrops came to sort out the riot, he was nowhere to be seen.

Our names were among those taken by the military police and, although nothing came of it, we mutually agreed that the bars of Sauchiehall Street were not for us. Every night brought several fights along its length, also up the side roads and alleys. The Scots fought the Americans, the navy fought everyone else and the Irish fought each other. The police, both civil and service, were none too gentle either. They used their truncheons freely, were reputed to slip on knuckle-dusters and believed in keeping the odds at least four to one in their favour. But then they were up against real streetfighters who knew the tricks of the trade. Weapons used included ice picks, screwdrivers ground to needle points, specially sharpened bicycle chains, bricks with barbed wire wrapped round them and the favoured local weapons — razors. The latter were produced within the time it takes to bat an eyelid. Another Glaswegian refinement was to sew razor blades into cloth caps for butting and swiping.

Back at the hotel we conferred. There must be some other way to spend our evenings. How better, said the manager, than to take advantage of what was on the noticeboard. Until then we had not bothered to read the board. It looked too official, being covered with green baize and crossed by white ribbons. Stuck in the ribbons however was a veritable cornucopia of ways to spend our Glasgow nights over Christmas and Hogmanay, now nearly upon us.

There were invitations to theatres, cinemas, concerts and dances. Invitations to eat at humble cafes and the best restaurants. Invitations to private houses and parties. These were for one, two or as many airmen who could make it. Everything was free and, in most cases, transport would be provided. Not that it was needed because, as the hotel manager informed us, serviceman could travel anywhere for nothing. The generosity of the Scots astonished us and made Jock smile with quiet pride.

For our first night out, Skip and Timber chose a meal at a good hotel, most of the others decided to visit a dancehall, while Fancy and I favoured a party given at a private house. The odd man out for that evening was Paddy who said he had a great idea of his own and went off by himself. In our case, the invitation card informed Fancy and me to telephone a certain number. I did so and a cheery Scotsman answered. Och aye, he was delighted to have us. Och aye, he would collect us himself. Och aye, he knew the hotel very well and would be outside in about a quarter of an hour.

Our host turned out to be a strapping Scotsman in his forties. He had hair greying about the temples, blue though somewhat bloodshot eyes and a face creased by constant laughing. His handshake was like a vice and his frequent smacks around our shoulders could be compared with blows from a drop hammer. Outside stood a big black car. I recall the hotel manager came out to receive several cardboard boxes which clinked significantly. There was more laughing and backslapping before our host told us to hop in and hang on. His advice was necessary because he drove like a madman.

We had a crazy ride across Glasgow, our host exuberantly sounding his horn and carving up whoever dared to remain in front of him. I should also mention that during the hair-raising journey he sat sideways with one hand resting on the back of the front passenger seat. I happened to be in the hot spot and received playful thumps whenever some point was emphasised. Fancy and I were also given a guided tour of Glasgow and informed about this building or that after he had raced past them. Civic centre architecture changed to residential and we were in a district with mansions standing well back from the road. Our driver seemed to be trying for the land speed record down the road, then braked all of a sudden and shot between two tall stone gateposts. Next he had to do an emergency stop because the drive was full of cars. It was a strange sight, I should add, as petrol was scarce in wartime.

Threading our way through the cars, we reached stone steps leading to an imposing front door which opened at a touch. Or rather the door started to open, then had to be pushed due to the weight of people inside. The hall, rooms, staircase and landings were thronged with revellers. I have never seen so many crowded into one house. What space remained was taken up by Christmas decorations, bunches of balloons, paper streamers, lashings of food and, most conspicuous of all, bottles. Everyone held glasses and seemed to have their own bottles. These bottles had lives of their own, emerging from cellars and passing from hand to hand, thereby ensuring not a single guest remained thirsty.

Our host introduced us to many of the partygoers and I ended up with a dark laughing lady in her late thirties wearing a low cut red dress and a broad gold wedding ring. She appeared to think me the wittiest person in the world for whatever I said she laughed appreciatively. There was dancing to a radiogram in one of the rooms and the laughing lady also demonstrated that I was the best partner in the world. It was like dancing with quicksilver. My feet never managed to tread on hers and she did not seem to have a bone in her sinuous body. As the night went on, we separated but came together again. Everyone was friendly. Many of the men wore evening clothes. One had the chain of some city office and others looked just as important. Fancy remarked on all these luxuries in wartime and wondered what the authorities would think. There was hearty laughter among the group he addressed and a jovial character said,

"We are the authorities."

The party began to break up about dawn. My dark lady said laughingly but pointedly it had been fun. She departed with her husband in a chauffeur-driven car, then one of the authorities took us back to our hotel. He informed us our host was a whisky distributor and that we should have helped ourselves to a few bottles as he had done. Would we like one each. I took one for Timber while Fancy, who had drunk too much already, had a bottle for himself. After our driver had dropped us off and departed, Fancy stood swaying outside the hotel. I managed to steer him to his room before going to mine.

There was barely two hours left before we would be roused to start another day's work. I decided to try and freshen up with a shower and had just picked up a towel and soap, when there came a tap on my door. Thinking it was Fancy, I told him to go away, but the tapping was repeated. Eventually I opened the door to see Paddy outside. His greatcoat had lost some buttons, his tie was undone and his trousers caked with mud. Instead of being fore and aft, his RAF cap lay athwartships.

"Thought you'd like to know," said Paddy in a slurred voice, "I've done it."

"Done what?" I had to ask, which meant he came into my room and stood swaying in the centre of it.

The story that followed was so ridiculous it has to be told. Apparently on one of our previous evenings out, Paddy had tried to buy a bottle of Scotch to send home to his father — "as a change from the Irish stuff, y'understand?" He was informed that whole bottles were not for sale, though one barman offered to sell Paddy an empty bottle, which he could fill up by the simple expedient of buying two tots at a time. This was the special mission our Irish colleague accomplished on his own. With the precious bottle in his pocket, he went from bar to bar ordering two tots — "one for meself and the other for me auld father, y'understand?" At last, after many adventures, mishaps and a slight argument with a Protestant, Paddy managed to fill the bottle. He then put on the cap and affixed some gold foil his barman friend had also sold him.

Smiling proudly, Paddy produced the bottle from behind his back. It was, he pointed out, a well known brand. The contents looked authentic enough, though who knows what went into their blending. At least his bottle was full to the very top and Paddy was especially proud of the way he had fitted the gold foil. He held out the bottle to show me what a good job he had made of it.

As Paddy did so, the bottle slipped out of the foil, hit the floor and shattered, filling my room with broken glass and the smell of whisky. It had been a long night and this was the last straw, but Paddy looked so disconsolate I decided against expressing my feelings. Giving him the bottle intended for Timber, I steered him out of my room and began the long task of cleaning up. An hour later the place still reeked of whisky, despite a wide open window letting in the raw Glasgow air. Perhaps that city had a similar aroma.

We worked through Christmas Eve but were given an hour off to do some shopping. Our purchases took the form of small presents to one another — a packet of cigarettes, a bar of chocolate or a cake of soap. These were self-consciously exchanged at breakfast on Christmas morning. No sooner had we finished the meal when the special police sergeant, who banged on our doors each morning, informed us of our Christmas Day duties. We were to put up decorations in the hotel dining room — first for our dinner, then for the party afterwards.

"There are the decorations," the SP sergeant pointed to boxes of them, "while you, you and you will volunteer for present wrapping."

I was with the wrappers and became intrigued by the fact that the gifts we had to cover up consisted of train sets, cuddly dolls and suchlike toys. Sparks and Taff who shared the task were equally intrigued but left me to question the formidable sergeant in charge of operations. Why were we wrapping up childrens toys? "Because lad it is going to be a childrens party. The RAF has invited all the kids in Gallowgate. The hotel will be full of them. By the time you go to bed tonight lad, you'll know you have had Christmas."

There is an old service tradition on Christmas Day, of officers and NCO's serving the other ranks. Thus around one o'clock, we were seated and singing "Why are we waiting?" As soon as we each had a plateful to be going on with, the cooks and waiters served themselves and sat among us. It was a boisterous meal with much leg as well as cracker pulling. Paper hats were put on, riddles read out and second helpings there for the taking. Flaming puddings were carried round the tables accompanied by the inevitable recitation about the grey-haired old veteran who had fought at Kyber Pass.

"You can take your Christmas pudding," we all chanted at the appropriate moment, "and stuff it up your arse."

Then we ate it. Beer, whisky and even liqueurs kept coming our way. Several people who indulged too much sloped off to their rooms. One actually slid under the table and we carried him, as rigid as a mummy, to his bed. These drop-outs reduced the number of volunteers for the kids party.

I shall never forget those kids. They began gathering outside the hotel at about three that Christmas afternoon. Most came by themselves, being more streetwise than we were, but the younger ones were brought by their mothers. We let the early arrivals in, because it was cold outside and at first they sat strangely quiet, though they noted everything. Master of ceremonies was the SP sergeant. While being wonderful to the kids, he bossed the rest of us unmercifully. We were each responsible for half a dozen children. Our duties included joining in all the games, which became increasingly hectic. We also had to serve food and clear up afterwards. In addition there were frequent toilet demands as the hotel facilities were far superior to anything in the surrounding homes.

Then came the performers. They included a clever conjurer, a crazy juggler and a clown who was the firm favourite. Someone played the piano so that the assembled company could sing carols and popular songs. And there was Father Christmas. The SP sergeant said he was going outside to hold the reindeer and not long afterwards a big fat Santa Claus came in with two enormous white bags on his back — one clearly marked Boys and the other Girls. Father Christmas had a kind word to say to each lad and lass, most of whom were wide-eyed and quite tongueless. An older boy ventured to say he thought he knew who Santa Claus was, whereupon the white-bearded, red-robed figure said a bit gruffly,

"Watch it, lad!"

Later, when the last kid had departed and the place tidied up, those of us who had managed to stay the course sat around the dying fire. The SP sergeant was next to me and he asked what part of the country I came from. When I replied not country but Wimbledon in South London, he went on to enquire if I knew Wimbledon Common?

"Yes," I replied, "I used to ride my bike all over it."

"But," he pointed out, "you are not supposed to ride bikes on any London common. Local bye-laws."

"I knew that," I nodded. "The funny part was they had a keeper to try and catch cyclists, but he was on foot. I used to give him an awful time."

"I thought I knew you," said the SP sergeant and burst out laughing like the rest of us.

We were back at work on Boxing Day. The objective was to get the merchant ship away before the end of the year though that proved impossible. Therefore it looked as if we would have to undergo the dreaded Hogmanay about which the Scots kept telling us. Christmas apparently was nothing up there. Hogmanay was the real festival where police were never called out — not even for mass murders — due to their own illegal activities. As the nights leading up to Hogmanay consisted of non-stop drinking and riotous behaviour, we began to view New Year's Eve with a certain amount of trepidation. Would we be called upon to tackle the monstrous haggis? Was it true that they served whisky with porridge? Sassenach rumours circulated and were supplemented by those from the Scots. We did not know what to expect.

As usual the unexpected happened, at least to our crew. About a couple of days before New Year's Eve, Skip mentioned that we had been invited to the festivities at Fiona's place. There was of course no obligation for any of us to go, but the skipper added in his usual understated way, Fiona tended to do these sort of things rather well. The uncertain consulted Timber who replied that we would all be going. So on the evening of the 31st December we returned from our labours to wash, put on best blue and await events.

At about seven-thirty an army vehicle drew up outside the hotel. The two men in the front looked exquisite with lace at their throats and wrists. One wore a kilt and the other tartan trews. They also had such accoutrements as feathered bonnets, sashes, daggers tucked in their socks and, in Sparks own words, "furry bags over their balls". Our Cockney was of the opinion they looked like a pair of Jock pansies until Skip advised,

"Better keep your voice down. Hamish and Alistair are both from crack Highland regiments."

We got in the back of the van and were driven northwards for over an hour. By then Glasgow was far behind us and we appeared to be crossing an interminably blasted heath. Eventually we turned off the road, passed a large lodge and continued for a mile or more along a drive. There were rock gardens on either side about two hundred feet high, after which we saw a castle perched on a mountain top and looking like a film set for Macbeth. The castle boasted turrets and battlements adorned with stone eagles, dragons and other nasties, each poised as if to swoop on us. The great iron-studded doors were wide open and lights poured down the steps with no heed for petty wartime restrictions. Littered about the forecourt were about half a hundred vehicles which caused our skipper to smile as he murmured,

"Good old Fiona."

Actually Fiona proved to be young, golden-haired and another of Skip's sensations. She wore traditional Scottish costume as did most people there. Her black court shoes had, I recall, silver buckles on them and Fancy kept nodding approvingly at such quality footwear.

How can one describe the magical twelve hours that followed? We wanted for nothing. There was warmth due to room-sized fireplaces filled with blazing logs. There were long tables laden with food and drink. There was friendliness from everyone — from gnarled old boys, who looked as if they had fought for Bonnie Prince Charlie, and from pert little maids, all flirting like mad. The dancing alternated between modern and traditional, and music was supplied by players who formed different groups while not dancing themselves. The ladies danced energetically, the men meticulously with inherent Scottish pride, toeing their way round swords laid on the dance floor. Of course there were the pipers who had us all galloping and the lone piper who led us slowly round the castle with the coming of the New Year.

We sang Auld Lang Syne and still the festivities continued. It was as if night, sleep and tiredness did not exist. We must have eaten half a dozen times while drinking was continuous — not hard drinking, but a constant topping up of glasses. At last the dark windows lightened and piles of ashes glowed in the great fireplaces. Breakfast appeared on silver trays for us to help ourselves and, only after that final feast, were we allowed to thank the still lovely Fiona and depart.

Fortunately no work was expected on New Year's Day. We slept at our hotel amid a silent, exhausted Glasgow. Next morning the tempo remained slow as the city struggled to life. Yet we were obliged to quicken our pace. The skipper of the merchant ship was fretting to leave and join a certain convoy. A day and a half later, all was done. The work was inspected and signed for. We cheered as the ship left dock with our eight air/sea rescue launches lined along the deck. Later we learned that the merchantman had been torpedoed in the Indian Ocean. Fortunately most lives were saved but our boats lost. By then we were well into the fateful year of 1944, a year in which everything that had gone before could be regarded as minor exercises.

RAF high speed launches 186, 137 and 176 in harbour but on standby duty with all three flags flying. (Courtesy R.A.F. Museum, Hendon).

Chapter 6

BUILD-UP

The turn of a year is as good a time as any to review one's life or, indeed, the progress of a world war. And where better to do so than on a slow train from Glasgow? In the air, the introduction of long-range fighters made it possible for our aircraft to hit just about any target in the Third Reich. At sea, the U-boat menace was ending, thereby allowing ships and supplies to pour across the Atlantic. Decisive victories however must be won on land and these too were forthcoming. The Russians were fighting and winning the greatest battles which resulted in the Wehrmacht being pushed back to from where it came. The Allies also enjoyed victories in North Africa and Italy. As a result of the latter campaign, the Italians dropped out of the war. I tell of these events to set the scene, also because they may not be familiar to present day readers. At the time they formed part of our daily lives.

Yet despite the encouraging trends I have related, there remained a long way to go, a supreme effort to be made, a lot more lives to be lost or in our case to be saved. The German war machine still worked with ruthless and terrifying efficiency. Although the Red Army had pushed the Wehrmacht back from the Volga to the Dnieper, that was 500 miles from Germany's original borders. As for Italy, the Italians may have surrendered, but stubborn German resistance continued. There was only one sure means to a decisive victory. A Second Front had to be established on the north coast of France then thrust eastwards into the heart of Germany. It was not going to be easy, due initially to the Atlantic Wall built by the Germans, followed by the enemy's reaction to any beach head the Allies might establish. Everyone realised this — from political leaders and military commanders to war workers and particularly those of us in the services. During 1943, work was carried out on the invasion plan code-named Overlord. Although the detailed plan was a most closely kept secret, we knew it would be implemented during 1944. The action was due to begin at what was called H-Hour of D-Day.

Such were our thoughts and conversations during the train journey. They were initiated by a remark Timber made.

"Yanks," he said, "I've never seen so many."

We had met them in Glasgow, but the numbers of Americans appeared to increase as we moved south. There were vast camps outside towns, beside roads and railway lines. Apart from living quarters, endless rows of trucks, tanks and guns lay over the fields. Yanks packed the stations through which we passed, also the train in which we were travelling. We found them friendly and talkative. We were having a few local difficulties, they informed us, so they had come over to help us out. Like last time.

"Soon lick those Nasties into shape for you."

"Great little country you've got yourself here."

"Your dames! Wow! Gonna take myself back one."

"But your railroads are so goddam lousy. Where I come from we've the Aitchison, Topeka and the Santa Fe. Y'know, like in the song."

The predominance of Americans became more apparent when we reached London. There we had to unload our packing cases from the Glasgow train, put them on RAF lorries, then cross the City to Waterloo. During that relatively short journey, we saw more Yanks than Brits. Waterloo Station was seething with them and each seemed to have acquired a female companion. The Americans wore fantastic uniforms covered in badges and medals. They also carried great quantities of equipment, food and drink. Bright-eyed girls seemed part of their kit. While waiting for a barrier to open, we watched an Anglo-American kiss that went on for the best part of ten minutes. The girl was trying to climb up the American who helped with some supportive work on her buttocks.

"Disgusting," said Fancy.

"Filthy but it's fun," Sparks grinned.

Back at Calshot, we had to unpack and get aboard our old boat. While we were doing so, Doc arrived in a station sickquarters car with Mrs Doc beside him. Apparently he had some momentous information. He could not tell us because Timber was fretting to reach our boat. What about in the canteen that evening? Seven o'clock sharp. Those of us who could go, promised to be there.

Doc's news did not seem momentous to me, though I nodded with the rest. The Yanks, he informed us, were coming down from the north. Great armies of them. Netley Hospital had already been taken over by the Americans which meant Mrs Doc no longer had a job there. Fortunately she was able to help out at Calshot though this too made problems. Rumour had it that all air/sea rescue boats and crews would be involved in the invasion operations. Most were to be posted away from Calshot with Doc once more having to leave his marriage partner. "C'est la guerre", someone remarked none too sympathetically because, in our opinion, Doc did all right. He had enjoyed himself at Calshot while we were toiling away in Glasgow. Anyway, the rest of us wanted to be at an air/sea rescue base. It was what we talked about while on various attachments. Mrs Doc must have sensed our reservations, because she started telling us about the Yanks at Netley Hospital.

"They've really taken over the place. Changed everything. You have no idea. Ice water machines everywhere. Patients fed great quantities of ice cream. Ice cream? And they're so loud, so crude. Their language! The women as well as the men. Also their medical staff are quite different to ours. With us, doctors are highly respected while surgeons are treated like gods. Their surgical teams are. . .well, clowns." Here, words failed Mrs. Doc. She had encountered situations later to be portrayed in television comedies such as Mash.

"Tell them about the admiral," her husband suggested.

"You tell them," Mrs Doc frowned in disgust, got up and left us.

"There was this admiral," Doc leaned forward in a conspiratorial way. "One of theirs, thank goodness. He had come in for a check-up. High blood pressure. A crabby old devil by every account, but an admiral all the same. The admiral was given a private room which he regarded as his quarterdeck. Ordered everybody about and upset the whole hospital."

"What's new?" Sparks pointedly enquired.

"It's what happened," said Doc. "One of their doctors thought it time to take the admiral down a peg or two. The patient had to have regular temperature readings and the doctor decreed this should be via the rectum. Next, one of his colleagues — fully masked — made the admiral lie on his side away from the glass panelled door. Then, do you know what he used instead of a thermometer. While half the hospital watched, he carefully inserted and left the admiral lying there sporting a large daffodil."

Our days were again occupied getting the boat fully operational. This meant having no faults or short falls anywhere. Each new crew member had to sign for equipment under his responsibility. In my case, an engineering officer went over the engine room with a fine tooth comb. It was the same on deck with the armaments and wireless. Particular attention was paid to direction finding gear and we had a series of simulated rescues. The exercises were based on the following routine.

When an aircraft came down in the sea and its crew managed to escape in a rubber dinghy, they used a small transmitter to send out a distress signal. This was located by our direction finding equipment which, when it picked up the signal, also gave us a compass bearing to follow. For our exercises, one of the craft would act as the dinghy while others worked as searchers. Later we used a real aircraft dinghy with a volunteer in it winding away at the transmitter. It was a cold, wet February day and I well remember his relief at being rescued from the bleak Channel.

Meanwhile notable changes were taking place throughout the Solent area. We were invariably criss-crossing it between Calshot, Southampton and Portsmouth; Hamble, Gosport and Cowes. We therefore saw more than most. Everywhere we looked — on all the beaches, up every river and creek — there were lines of landing craft. More and more kept arriving in the weeks that followed. Timber declared they were not proper boats but I was fascinated by them. They came in all shapes and sizes, many after travelling across the Atlantic. Mostly they were painted a drab grey with big black letters designating their purpose. For example I learned that LCI means landing craft infantry, LCT landing craft tank and LCM landing craft to take mechanised vehicles such as supply trucks and self-propelled guns. There were also LST's or landing ships capable of carrying several great battletanks. These had the usual type of ship's bows which would swing open on arrival at the invasion beach like a pair of colossal doors. Then the tanks would come lumbering out, one after the other, in a fearsome show of armoured strength.

As spring and milder weather approached, I would often take a small boat on visits to these landing craft. By then there were rows of them moored everywhere with bored looking sailors leaning over the sides. Visitors were welcome to share a brew of tea and a yarn or two. One crew, who had sailed their craft from the States, told me of another which had vanished one night in mid-Atlantic. I also learned they took their vessels in turns to Slapton Sands, South Devon for training exercises. All civilians had been evacuated from the district. This became thoroughly battered as wave after wave of landing craft ground ashore. Then combat troops firing live ammunition and hurling real bombs rampaged around what was left of the family homes.

I once saw a number of landing craft to which intricate superstructures were being welded. These consisted of frames holding hundreds of forward leaning tubes.

"Nine hundred in total," an armaments officer informed me. "They take rockets. Think of it. Nine hundred rockets, with high explosive warheads, being fired simultaneously. At the touch of a button." I did not like to think of it, but we were at war.

By then guns bristled everywhere one looked — from the huge guns of battleships assembling at Spithead to multiple cannon and machine guns being mounted on all boats. Our rescue launch was no exception. We had to have newer, better guns. The armaments people swarmed over us, then we put to sea for target practice. There were target rafts to hit while racing at top speed, also an aircraft towing another target round and round us. Our armaments instructors enthused about armour piercing and high explosive shells, alternating with tracers which showed the direction of fire.

Back at RAF Calshot, other experts in death and destruction came to lecture us. They were mainly officers covering the complicated aspects of amphibious warfare — American as well as our own RN, Army and RAF personnel. The detailed information imparted to us has become blurred with the passage of time. So have the faces. Yet, for some reason, two lecturers and what they said remain lodged in my memory — one perhaps for the clarity and unexpected passion with which he presented his material, the other for his enthusiasm.

The former was the only civilian who addressed us. He came bowler-hatted and carrying a rolled-up umbrella. Later someone remarked he was the kind of person who would pass unnoticed in any city street. This unlikely lecturer had a dry fluting voice which caused a few backrow titters in the service audience. He must have been high up some scale — a professor or a senior civil servant — for our commanding officer no less was present and treated the newcomer with the utmost deference. The subject, which sounded dull before the speaker began, was on planning an invasion.

Planning for the invasion of Europe, we were told, had begun with a series of mathematical calculations to try and answer vital questions. Where were the places for our attacks to be launched? What were their relative advantages and disadvantages? How many men would be needed for the initial attacks and how many more for the subsequent follow-ups? How many boats were required to carry them? What weapons should they take? What supplies of ammunition, food, fuel and other items of equipment would they need to fight their way ashore? The little lecturer reeled off hundreds of facts and figures without referring to a single note. He was particularly knowledgeable on enemy dispositions. We were told exactly how many troops the Germans maintained in Northern France. The numbers were amazingly high despite the Wehrmacht's Russian, Italian and South East European commitments. It seemed impossible for us to take them on in a straight fight, let alone begin this by crossing the Channel and trying to pierce such formidable defences.

At least three million men, we were told, would be needed to overcome the enemy's western divisions. To convey such a number to France required thousands of landing craft, protected by hundreds of warships and provided with overwhelming air power. Naval escorts would range from battlecruisers to the motor gun boats we knew so well. Air cover would entail the combined strengths of British and American air forces. Operations were envisaged that would make recent 1000-bomber raids seem like side shows. Our aircraft were being briefed to hit everything German, also whatever else served to help them. That meant roads and railways, bridges and canals, as well as all their military installations. In addition to aircraft bombing and strafing, hundreds of gliders packed with combat troops would be towed to and released over France. Tens of thousands of paratroopers were due to be dropped on enemy-held territory. It was to be the largest such operation ever carried out in the history of warfare.

Point by point, this mild-looking man prepared us for Armageddon. Already, he told us, the three million Allied combatants were being assembled throughout the length and breadth of Britain. More than half would be Americans. British and Canadian forces together totalled around one million and, in addition, there were substantial French, Polish, Czech, Belgian, Norwegian and Dutch contingents. At this stage, as the lecturer neared the end of his address to us, he changed his tone.

"You are involved," he said earnestly, "in a crusade. You should remember that the Nazis plunged the world into the bloodiest of wars. They placed 300 million people in bondage. Millions have been taken from their homes and from their countries to work as slave labourers. Millions more have disappeared into camps where torture and death are all too common. Your crusade is necessary to end this era of savagery, to prevent a slide into barbarity unsurpassed in human history."

We were impressed because Korea, Vietnam, Lebanon and the rest lay in the distant, unknown future. Despite being a hard bitten service audience, we clapped the civilian lecturer. One more effort, most of us thought, and that would be the end of man's inhumanity to man.

The second speaker a few days later was the enthusiast I mentioned. He was a hefty, ginger-moustached tank specialist. Until then I had always thought of tanks as awkward lumbering vehicles capable of crossing rough terrain and firing guns. I knew from war reports that the Germans had used their armoured divisions with what seemed like unbeatable might — until the Russians smashed them. Meanwhile the Americans, employing mass production techniques, were turning out tanks in prodigious numbers. In the face of these formidable achievements for armoured warfare, what else could the British add? Our tank enthusiast was there to tell us with the aid of photographic enlargements.

"Apart from this conventional type of battletank," he held up his first picture, "here you can see what we call in the trade a bulldozer tank. It is adapted to clear beach obstacles. They are simply pushed aside, exploded or ground into the sand. The next variant is a flail tank. Those chains on the front rollers are there to beat the ground and set off whatever mines the enemy has left in its way. It can fire landmines without damage to itself. Now we have a tank which puts down a track while travelling over soft ground for other tanks to follow. We also have various kinds of swimming tanks and the special one here is a submarine tank. Which brings me to tank horrors — flame throwing tanks and tanks that hurl high explosives. My two favourites come next. The first is a bridge-building tank. Those structures carried on top unfold to span a fair-sized river. The second is what might look to you like a turretless tank. It is in effect a self-propelled ramp which is driven up to, say, a sea wall. After that the other tanks travel over it to get to the top of the wall."

The tankman went on and on mesmerising us. I was to recall his approach long after the war when I worked, strange to relate, in Southern Germany. I headed a small team of English engineers on a large German project. Both nationalities pulled their weight, but the British possessed an advantage which the Germans recognised.

"We haf to haf a plan," was how the German bossman put it. "If our plan it does not work, then ve haf to haf another plan. But you English. . .warten sie nicht langer, you do not hesitate. You. . .versuchen, erfinden. . .improvise. You English are masters of improvisation."

There were many notable improvisations to be seen about the Solent in that spring of 1944. From Calshot, we watched giant steel and concrete structures rising along the far shore. At first we had no idea what they were but, so many people worked on them, it was easy for us servicemen to be given guided tours. Timber had a friend at Gosport who used to drink in the evening with a high up civil engineer. We went across to see the friend and it led to the engineer telling us what he was building. He did so in one word.

"Mulberries."

It came out he was referring to the basic building blocks of the now famous Mulberry Harbours. They comprised Mulberry A and Mulberry B, each the size of Dover Harbour, which is massive if one stands back and looks at it. These prefabricated and portable harbours were towed to France and used until the Allies were able to capture permanent ports.

When we first saw them, the Mulberries were concrete caissons that somehow miraculously floated. Steel framework was added to those floating foundations and encased in more concrete. Eventually the structures looked like city blocks without doors or windows. Deep within them were quarters for harbour personnel and anti-aircraft gun crews. Deeper still lay the ammunition stores and electric power generating sets.

As we made our way along catwalks and dizzy heights, the civil engineer explained how the giant building blocks would be sunk side by side to provide an inner breakwater. He went on to say there would be outer breakwaters of steel floats and blockships, filled with yet more concrete to protect the prefabricated harbours from the full force of the sea. Fleets of tugs from all over the British Isles were being assembled for the towing operations. Everyone concerned hoped the weather would remain fine because even the simplest towing job needed calm conditions.

Other interesting sights we inspected were what looked like huge floating cotton reels. These formed part of the Pluto (pipe line under the ocean) project by which the vast quantities of fuel required for tanks and trucks would be pumped from England to France. The object of Mulberry, Pluto and other similar enterprises was to build up men and machines, arms and equipment in the bridgeheads faster than the enemy could move against them. As the engineer succinctly put it,

"They have to work or all will fail."

In his play Henry V, Shakespeare used a telling phrase before the Battle of Agincourt — "Now thrive the armourers". It was exactly like that along the South Coast in the spring of '44. During my cycle rides about the vicinity, along the Solent and into the New Forest, I constantly heard the sounds of metal striking metal. Occupying one field would be dozens of trucks having armour plating fitted to protect their driving cabs. Another meadow contained hundreds of guns being assembled. Civilians worked with servicemen. Girls in khaki and blue dungarees toiled like navvies. Roads had to be adapted. Hedges were hacked down, corners cut back and long lay-bys established. New entrances were made into roadside fields, then through a succession of meadows with gateposts removed here and hardcore laid on soft ground there. Tented cities sprang up overnight. There were tents for sleeping quarters, tents for field kitchens, tents with red crosses, tents firmly closed and closely guarded. Railwaymen were equally busy, laying extra tracks to carry increased weights of traffic. Over 200 miles of rails were added to the existing network in order to take military supplies to southern ports. As for airfields and airstrips, about 150 new bases were constructed. Pilots claimed they could practically taxi along the coast from Dover to Lands End.

During this build-up period, I saw both Eisenhower and Montgomery. I became used to seeing Ike because he lived in a house along the road from my Wimbledon home. He used to whirl past our front gate, usually driven by a lovely-looking girl, with military police and his own staff officers in close attendance. Once, when I was near Havant helping Sparks to collect more mysterious wireless apparatus, we heard that Eisenhower was about to address allied troops in a nearby field. We hurried there and waited. Presently his car and outriders rushed into the field and the next moment there he was on a small platform fitted with a battery of microphones. The mikes were connected to loudspeakers lodged in nearby trees but a South Coast breeze also became involved. The great man's words were swept away and we only captured gems like "great" and "sure am". But we cheered and Ike waved before rushing off to his next appointment. I thought that, behind the famous smile, he looked worried — as well he might be bearing the top responsibility for three million lives.

Montgomery was altogether different. He preferred gatherings of a size to make his prissy but quite pleasant voice heard. Any object which raised him up a foot or two sufficed for a platform. Also, he disdained microphones. Once Monty had our attention he went on to capture our interest — talking about the war in cricketing terms. We had to bowl fast balls, keep a straight bat and knock the enemy for six. It worked because, as soldiers who fought under him told us, Montgomery meant what he said and did not throw away lives.

Since the beginning of 1944, we had remained at Calshot. We were kept busy because the RAF did not encourage idleness, but we took no part in any operations. Nevertheless our craft were in tip-top condition, we lived wholly aboard them and we stood by to go anywhere at a moment's notice. In fact that was the experience of most crews being posted. Their skipper would be summoned ashore and not long afterwards they were on their way. As these launches departed, crew members would call out places such as Ramsgate, Poole or Salcombe. One afternoon the launch with Erik the husky on board gathered speed past us. Some wit had strapped a tin hat to the dog's head to signify he too was going to war. Erik just grinned happily.

Doc was not happy to be brought back aboard as a member of the crew. He kept moaning about the move being unnecessary. He could have joined us within minutes from station sickquarters. Now he had to hang about, worrying all the while about his wife. If only the authorities would reveal where we were going, then Mrs Doc could make the change as well. Someone mumbled something about security but Doc would have none of this. He went over his old arguments about medical orderlies and nurses being different. Even if we eventually went to France, Doc told us, nurses would be required over there. I recall Fancy casting his eyes upwards and Timber, busy with yet another letter home, sighing. At that moment, Sparks called down from his wireless cabin,

"They want the skipper ashore. I think this is it."

Skip went ashore and was away some twenty minutes. The rest of us hung about the deck keeping our eyes on the dinghy he had left moored by the pierhead. Presently we saw him returning along the pier. Our skipper went down to the boat, started its engine and cast off. After that, all eyes were on the bow waves of the dinghy, creaming away on either side as the boat headed straight for us. Skip came expertly alongside, we helped him aboard and he gave us the news as follows.

"Weymouth. No. 40 Air/Sea Rescue Base. We'll be one of eight boats there. It's a definite posting this time. We leave right away, but I sent word through to your wife, Doc."

That last little touch was typical of our skipper. He would hear our various problems and, if there was anything he could do, Skip quietly did it. But now Timber was issuing orders. "Hoist the dinghy aboard. Clear the decks. Start the engines. Put on your inflatable jackets". We scurried in all directions. Fancy and Paddy lashed down the dinghy. Taff hoisted the RAF ensign. The skipper and Timber conferred over charts. Jock and I warmed the engines. Like Erik, we were off to war.

It was a beautiful spring morning, the waters of the Solent sparkling as we gathered speed past Calshot Castle. An aldis lamp winked "Good luck" to us and we flashed back "Thanks". Then we turned by the red lightship and set a course which took us between the New Forest shore and the Isle of Wight, passing Lymington and Yarmouth, Hurst Castle and the Needles. Out in Poole Bay, we opened up and our boat began planing perfectly on the calm sea. We sat on deck talking excitedly. What would No. 40 ASR hold for us? Would we come under the navy as usual? There was we knew a large naval base at Portland. We had no idea as yet how different the Weymouth posting was going to be. For the time being however, we were caught up in the sheer exhileration of that short journey. From Hengistbury we made for Durlston causing Bournemouth and Poole to sink away on our starboard. Then, after St Alban's Head came that wonderful line of white Dorset cliffs marching westwards by Purbeck, Lulworth and Chaldon. Ahead, like a finger pointing out to sea was Portland Bill. The massive sides of Portland Harbour came into view as we entered Weymouth Bay. The town was also visible but something there was clearly wrong. While the cliffs shone white and the bay remained bright blue, a dense cloud of black smoke hung over the vicinity.

We cut our speed to dead slow on the approach to the entrance of Weymouth Harbour, wondering all the while about the black cloud. Across the bay, nestling under the cliffs by Osmington, were the usual lines of landing craft. Nearer to us lay some American destroyers from behind which shot a couple of patrol torpedo boats that made straight for us. The PT boats looked as if they were either going to ram us or shoot us to bits. Only at the last moment did each launch swerve aside. We caught glimpses of American sailors, their jaws chewing rhythmically, then we were left rocking in their combined wakes. We still had to try and figure out the strange black cloud over Weymouth.

We were given a clue at the harbour entrance. At the end of the breakwater stood a stovelike contraption from which black smoke poured. Stoking it, with what looked like oily rags, was a big negro. At first we thought him naked, but the effect was due to his overalls being as black as his face and hands. His eyes rolled and his teeth flashed when we went past him. By then we could see other incinerators perched on various vantage points and attended by similar stokers, all adding to the cloud over Weymouth. It was a form of smoke screen which the Americans used during troop movements. The townspeople hated it because everything became so dirty, but the whole area was on a war footing, Yankee style, as we were about to find out.

Between the harbour entrance and the town bridge, there is a fairly wide pool in which craft can usually manoeuvre. However we found this space crammed with dozens of PT boats. They lay six and eight abreast from the quay sides leaving a much restricted area in between. Our skipper wanted us to turn, so that we ended up facing the harbour entrance in true crash call position. Somehow Timber managed it — watched by hundreds of chewing American sailors. I was in the engine room for most of these prolonged manoeuvres and did not emerge until we were moored next to other RAF craft. We were directly outside the Customs House that served as wartime headquarters for No. 40 Air/Sea Rescue Base, Weymouth. A few seconds later, all hell broke loose.

We heard a plane engine droning above the black cloud, then every American in the vicinity opened fire with every available gun. Somehow word had got around that the aircraft was a German one, quite erroneously as we later learned. The Americans nevertheless fired off all they had — anti-aircraft guns, machine guns, ships cannon, automatic rifles and even revolvers. We could see GI's and sailors leaping about and shouting hysterically as they filled the sky with metal. None of them seemed to consider the effect of gravity. What went up had to come down somewhere. That was why Timber ordered us to keep under cover and this we did. Sparks however was nearest to the hatchway and, while not putting his head out, he placed a hand on the combing.

All of a sudden, Sparks gave a shout and came staggering back among us. He was staring unbelievingly at his right wrist which appeared almost severed and was pumping blood. I must say Doc acted swifter than the rest of us. While we stared, he pressed a piece of gauze and cottonwool padding on the wound, then wrapped it tightly in a towelling type of bandage. But still the blood kept coming through and Timber said we must get Sparks to hospital as soon as possible.

Without thinking, I jumped on the dockside and ran to the base building. The gunfire was nearly over by that time, but I was more concerned about finding a telephone. Luckily the first person I saw was a switchboard operator who acted at once. She listened to me, remarked that a hospital car would be quickest and made the necessary call. By then word was going around about the plane being one of ours though most Americans kept repeating how they nearly got that son-of-a-bitch.

We were more anxious for Sparks who was looking decidedly pale. To our relief we saw a civilian Morris saloon car, driven by a lady in nurse's uniform, trying to negotiate the crowded quay. She would have reached us much sooner, but the Yanks insisted on getting in her way. At last Sparks was placed in the car with a blanket round him and Doc by his side. The driver said it was not far to the hospital. She was an extraordinarily good looking woman in her early thirties. She had black hair parted in the middle and pulled back, large violet coloured eyes, a classic oval face and a small rosebud mouth. After Sparks had departed with her, we commented on it being typical of him to have such a glamorous driver.

Sparks was away overnight. At the hospital they found the wound was not serious, though it needed quite a bit of patching up. When he returned next morning, his right arm was bandaged from the base of his fingers almost to his elbow. The first move Sparks made was to check on his morse speed. Despite taking a cynical view of the war and all service activities, Sparks clearly did not want to leave our crew. After assuring himself and Timber he could still continue working, Sparks sat back and parried our questions about the attractive angel of mercy. He told us little, but more was to come out later.

Meanwhile, much had happened during his overnight absence for medical treatment. Being a proper base, the RAF at Weymouth had a large billet for its crews and we were taken to it. Our quarters were in a mansion which had once belonged to General Booth of Salvation Army fame. Everything had been removed except for a grand piano left in one corner of the main reception room downstairs. Upstairs, two or three bunk beds were put into each available room and there were two good bathrooms. Waafs came and cooked for us, also there was a similar large house next door containing ATS girls who manned anti-aircraft batteries overlooking Portland Harbour.

There were two routes from the house to the harbour. The shortest and most interesting way was via a series of alleys through a district that had been old in Nelson's days. These alleyways gave unexpected views of Weymouth Bay, Nothe Castle and the harbour itself. Timber loved the place and read me the description he put in a letter to his wife. He also told her and me there was supposed to be a lady ghost haunting the alleys. The apparition was named the Grey Lady and local legend maintained she waited for a naval lover who had sailed away in a ship of the line never to return. We used to go up and down those alleyways at all hours of the day and night but saw nothing. In the half light however, especially when sea mist filled the alleys, it was not difficult to visualise the Grey Lady still waiting.

After settling in at the billet, we wandered back to the harbour, then through Weymouth Town to the front. Nowadays it is a typical seaside resort but May 1944 was very different. The main differences stemmed from the Americans in their tens of thousands milling around everywhere. All the hotels, guest houses and private houses were filled to overflowing by Yanks. They were the lucky ones for outside Weymouth — towards Chickerell, Upwey and over the Dorset Hills — stretched tented cities. Every other US serviceman seemed to possess a jeep or truck. Beside the hotels on the esplanade stood long lines of jeeps forming a transport pool. The Americans simply left jeeps where they could on arrival and took the nearest when departing. Jeeps and great trucks were also ranged along the centres of most streets. Citizens who managed to remain in their homes said the place had become like a Wild West town.

Returning from the front we found a canteen beside the harbour bridge. Some good ladies of Weymouth ran it and, being a tea and buns sort of establishment, it was not favoured by the Yanks who preferred coffee and doughnuts. Immediately we sat down, Doc brought up his usual problem of getting Mrs Doc to join him. He had already made enquiries. A security clampdown restricted people entering or leaving the area. Even service personnel needed special permission and civilians were told to keep out. We informed Doc he could not possibly succeed this time. There was no room for personal considerations let alone cozy conjugals. The greatest amphibious operation ever planned had to take priority.

"The balloon," said Timber, "is about to go up." Our first coxswain finished his tea before addressing Doc further. "And another thing, you'll regard yourself as a full time member of my crew from now onwards. No swanning off to station sickquarters as you used to do at Calshot. I intend to run a ship that is one hundred percent operational."

But best intentions are not often kept. Next morning, that is when Sparks returned to join us, a message came through from our parent station. This was RAF Warmwell, quite a way inland. Station sickquarters there would like to see Sparks accompanied by our medical orderly. This could be done when the ration truck for the base made its daily visit to Warmwell. Someone from our boat had to go with the food waggon — to do the real work as Timber put it — and I volunteered for this duty. It would give me a chance to see beyond the coastline.

It was another beautiful spring morning. So far the weather had remained settled — perfect, according to the pundits, for the forthcoming invasion. Certainly the Dorset countryside was as lovely as I have ever seen it despite the area ten miles back from the coast being packed with combat troops. RAF Warmwell was also ready for action. Planes had been fitted in just about everywhere with mechanics working all over them. Occasionally an engine would start up with a powerful roar.

While Sparks and Doc visited station sickquarters, I picked up the rations for No. 40 base. These consisted of meat, vegetables and bread; butter, sugar and tea. The rest we were supposed to obtain locally in and around Weymouth. Whoever made that rule had no idea of conditions in the town. The rations were packed ready for me to take and I did not have to wait long. A much longer wait took place outside sickquarters as the medical officers there examined Spark's wrist. Naturally, while he was there, Doc tried to further his own ends.

"It's no good," he dolefully informed me on his return. "The security clampdown seems complete. No one can enter the coastal area, not even my wife who is a qualified nurse. Only a high ranking officer can authorise clearance and who do I know? We have never been separated like this. Never."

"How are you, Sparks?" I asked the real casualty.

"They're not too keen about me carrying on straightaway," our wireless operator eased himself into the truck seat, "but as I told them, I don't want to let my crew down. Perhaps," he gave a crooked smile, "they'll mention me in despatches. Give me a medal. For courage in the face of our allies. What did you think of Sarah?"

"Who?"

"The lady who drove me to the Weymouth hospital. Did you know Sarah lives along the road from our billet. Her husband's away in the commandos so she does part time hospital work. Sarah has invited me round to her place this evening — if I can make it. She mentioned some electrical work needed in the house. I'm sure I'll be able to help her. After all, one good turn deserves another."

Sparks talked about the divine Sarah all the way back to Weymouth and Doc about Mrs Doc.

Later that day we had to take our boat round and into Portland Dockyard for refuelling. It was a traditional naval base, still occupied by British sailors and Wrens, though they were a beleagured contingent entirely surrounded by Americans. The Yanks had put up temporary offices, workshops and stores which far exceeded the Royal Navy's efforts over two hundred years. The brightly lit modern offices, fully equipped machine shops and well stocked stores run by the Americans were going to prove a great help to us.

When I went to get some materials for our engine room, the RN stores did not want to know, as I had been warned back at base, so I tried the US Navy. Without any preliminaries I was taken to see their engineering captain who was having coffee and doughnuts and insisted I join him. I sat there a little awkwardly at first because in our navy captains were like celestial beings. The American captain said,

"You may dunk."

He was of course referring to the delightful habit of soaking the doughnut in one's coffee before biting off the end. After that offbeat opening, we were like old friends. There was however an interruption which I shall describe because it showed the American approach to most matters.

We were still dunking our doughnuts when two Yankee sailors shambled through the open doorway. One was a bull-necked petty officer and the other an ordinary seaman twisting a cap in his hands.

"Hell!" cried the captain. "Don't tell me Hagerty's in trouble again."

"He sure is," the petty officer confirmed.

"What is it this time?"

"First, he takes one of our trucks to see a dame at Dorchester. There he meets up with some old drinking pals. They do the bars and tangle with the snowdrops but Hagerty he gets away. He shakes off the posse by using the side roads. Well captain, you know what these goddam English roads are like?"

"He didn't. . .?"

"Yeah, he wrecks another truck, also brings down some telegraph poles and wires."

I was fascinated. In the case of our navy, there would be long drawn out proceedings during which po-faced officers would throw the book at the culprit. Perhaps it was the same in the US Navy, but I doubted it. Not then at any rate. I waited wondering what would happen next.

"Hell," said the captain once more. "Hegarty's so low down the scale, I can't bust him any lower. But I can bust you, Chief, if you ever let him anywhere near a goddam truck again. Case dismissed."

After they had left, the engineering captain told me he came from Hartford, Connecticut and said I should call him Mort. "If there is anything else you guys want," he stressed, "just ask for Mort". On our way back to Weymouth Harbour, I related the incident, though no one really believed me. That afternoon, as we were finishing our work for the day, a jeep drew up on the quayside and Mort put in a second appearance. The US captain said he was returning to his hotel and thought,

"Jeez, those guys could do with some honest-to-goodness American food."

While we stared open-mouthed, Mort proceeded to unload king-size cans of fruit salad, monster boxes of cookies and literally a crate of doughnuts. These represented unhoped for delights to us and we were duly grateful. Mort was invited below to meet the rest of the crew. Within seconds he was chatting away and, after he had departed, the skipper said,

"Delightful fellow."

On Sunday morning, Mort arrived once more. He spoke of taking a drive round little old Dorset and enquired who would like to accompany him. He had a jeep, but could easily get something bigger if the whole crew wished to enjoy the trip. We were neither on crash call nor standby so the skipper said three of us could go. He suggested me as I had made the first contact with Mort and, while pausing to think of who else, Doc volunteered. It was then decided that Fancy should join us. We had no idea what the US captain had planned.

It was another perfect May morning, the air so clear we could see great distances. Steadily we climbed the Dorset hills and, on reaching a vantage point, Mort stopped the jeep so we might enjoy the view. But we saw less of the view and more of the military vehicles moving bumper to bumper towards the coast. The main body of the US Fifth Army was arriving. There were battletanks, armoured cars, personnel carriers, self-propelled guns, the ubiquitous jeeps and long columns of ambulances. Engines whined in low gear and I can still hear the rackety sounds of tank tracks. In the villages through which these seemingly endless columns passed, the inhabitants stood and stared as we did. We were seeing sights no one had previously witnessed and, it was hoped, would never be repeated in the heart of the English countryside.

So we combined seeing "little old Dorset" with watching the invasion build-up. We went to Hardy's Monument, also to neolithic Maiden Castle above Dorchester. After that we visited an American camp city occupied, they told us, by over ten thousand GI's. Their bunk beds stretched into the far distance. Food mountains rose beside field kitchens. A catering officer said the mounds of food had to be rebuilt after each meal, also the chow queues were hundreds of yards in length. At the other end of the digestive process, latrine ditches were dug down wind with boards fixed above them. It was a rough place and we were glad to get out of it.

In contrast, Mort said, he was taking us to another place for our lunch. This turned out to be an ancient fishing shed decorated with genuine nets and glass floats. At the various tables sat high ranking US officers with the prettiest girls in the district. I recognised an attractive ATS lieutenant from the house next to our billets and she gave me the faintest of winks. Then giant menus were handed to us as if there was no war on. When we hesitated, Mort said,

"It's my treat, gentlemen. I can recommend the lobsters."

Afterwards we decided to try to walk down that splendidly rich meal. Mort and Doc strode ahead while Fancy and I followed at a more leizurely pace. Our path lay along cliff tops with the sea on one side and a deep valley on the other. The sea scene showed Weymouth Bay from a different viewpoint. Landing craft and destroyers lay below us like toy boats in a bath. Looking the other way, the valley was filled with military toys — tiny tanks and little guns.

That Sunday was both wonderful and thought provoking. Doc seemed to get on exceedingly well with the captain from Hartford, Connecticutt, leaving Fancy and I content to bring up the rear. Our drive back to Weymouth was not so easy as we had to join the slow tide of men and vehicles also heading in that direction. In the end we made it, observing on the way that large landing ships were being run up on the beach to take tanks and trucks aboard. While in the traffic snarl-up, we listened to the shouting and swearing which seemed necessary for such exercises. Eventually we reached the harbour, thanked and said good night to Mort. Doc was not only profuse. He took the US captain by both hands and pumped them vigorously.

No reason for this came into our heads at the time, but Fancy and I should have guessed. On Tuesday morning, as I handed Jock equipment for the engine room, Mrs Doc came smiling along the quayside.

"Is my husband around. . .?" she started to enquire, then Doc bounded up from the sick bay two steps at a time. They kissed and hugged before walking a little way from the boat. After Mrs Doc had departed, our medical orderly returned to a barrage of questions. How had he managed it? What about the security clampdown? Where would she be working? Doc kept smiling as we interrogated him. Eventually he said,

"Mort kindly managed it for me. Pulled strings in high places. He's a full captain, y'know. Aren't the Americans wonderful people?"

Chapter 7

D—DAY

We were now within two weeks of D-Day, but none of us knew the precise timing. Years later an ex-brigadier, who had been at Southwick House near Portsmouth, told me of the big room there where they planned and plotted the invasion. One entire wall, he said, was covered by a map of the Channel showing the south coast of England and the north coast of France. The map was so tall that the plotters worked on wheeled stepladders. Staff officers, such as my friend, were in direct telephone communication with various command centres. Thus he had an overall view. Our view was of a tightly packed town and hinterland, made even tighter as yet more troops poured into it. Emphasis was on the forthcoming conflict. Men were tense because they faced mutilation or death. They did not behave like book characters or movie actors. This, as one American put it to me, was for real.

What did for real mean? It meant many nasty scenes when discipline snapped and order had to be restored. There was a small army of military police keeping control by methods I have not seen before or since. As we had witnessed in Glasgow, our service police were hard but their transatlantic counterparts — the Snowdrops — were meaner and rougher. Two examples from many will suffice. There is an ancient inn by Weymouth Harbour where today's businessmen and holidaymakers enjoy themselves in pleasant surroundings. In the fortnight before D-Day, it was solid with GI's shouting, drinking, swearing and fighting. Once again, as we had seen in Glasgow, a big guy pushed his way through the crowd and demanded to be served. Pronto, he said. He was a huge man with, we judged, Indian or Mexican blood. The giant GI had had plenty to drink and a frightened barman tried to avoid serving him. With one ham-like hand the American grasped the bartender and the other hand produced a Bowie knife.

"You'll serve me, bud," said the GI, "or else".

"Yes sir," said the barman.

As before we were sitting well back. Already there had been a couple of fights and plenty of slanging matches, but the knife episode worried us. Who knew what the huge swarthy GI would do next. Actually he was drinking quietly when the Snowdrops put in an appearance. The bar door was flung open in best western tradition and a tubby little sergeant entered the bar which instantly fell silent. People parted, again as in a cowboy film, so there was no one between the small Snowdrop and the big GI. Then came a surprise. Evidently the two men knew one another.

"Ah've come to get you, Huey," announced the Snowdrop sergeant, hands on gun belt.

"You'll have to take me first, Fatso," replied the GI sticking the point of his knife into the counter top so that it quivered beside his drink. Paddy whispered to me,

"This is going to be interesting."

But the Snowdrop remained master of the situation. Calmly he asked everyone to move back still further so there was plenty of room. "Right back, folks", he said this several times. People did so and some scenting trouble slipped out of the door. The recently crowded and noisy bar room was no longer packed tight, no longer full of sound. Instead there was a new element, that of acute tension. "Move right back, folks", the sergeant kept saying. Then, without any warning, he pulled a gun out of his belt holster and shot the GI in the leg. The big man stared unbelievingly, first at the sergeant, next at his leg, before crashing to the floor. As he fell, three other Snowdrops rushed into the bar. They raced forward, grabbed the wounded man none too gently and dragged him outside.

"It's all over, folks," the sergeant informed us. "Enjoy yourselves, but watch it!"

We decided to leave because the gunning down of a drunken man was a sobering experience. An English policeman, Fancy remarked, would have handled it a better way. This was an extreme case yet savage fights, near riots and mass arrests were frequent during that uneasy fortnight. Another memorable event occurred the following afternoon.

Four of us — Paddy, Taff, Jock and I — decided to walk along the front. We automatically went about in threes or fours to try and avoid trouble. Suddenly we noticed Americans running. They shouted wildly to each other as they turned down a road leading to the inner harbour. Curiosity got the better of us and we took the same turning. A hundred yards away, a mob had formed and there were watchers at the surrounding windows. We asked a GI what was happening and he gave a spooky smile.

"We're going to lynch ourselves a nigger boy," he said. "He gone raped one of your white girls so he gotta learn he can't do that over here, like in the States. No suh."

Now he had told us, we could see more clearly. Struggling in the centre of the crowd was a tall, thin coloured man. Several whites held him while those who could rained blows on their intended victim. By then someone had obtained a rope from one of the nearby boats and the man was being dragged towards a lamp post. Our minds told us this could not happen in an English street, yet our eyes showed us it was about to take place. There also seemed nothing we could do about it.

While we stood gaping and undecided, an RN captain appeared by our side. He was a typical senior naval officer — trained to command and accustomed to be obeyed. He had cold, grey eyes and walked with a gold-headed cane.

"What on earth is going on here?" the captain asked us and, when we told him, he ordered us to fall in four square. I must confess we were reluctant to do so, but no one disobeyed an RN captain. So we formed up and this seemed to have an effect because the lynch mob parted. It gave us a full view of the terrified captive covered with bruises, blood and spittle. One American sailor had already formed a noose on the rope. This sailor looked sheepish when he saw our captain but the GI's holding their intended victim were unimpressed.

"He raped a white girl. . ." one of them started to say.

"That is why I am arresting him," the captain coldly interrupted. "Escort," he said to us, "take charge of your prisoner."

I do not know who was more terrified, the coloured man or us. His captors must have loosened their grip for the next moment he had slipped into our little square of four. Immediately the captain called us to attention, then gave the order to quick march. There were some mutterings among the crowd, also one or two shouts from the back about American justice, though that was all. To this day I do not know how we got away with it — or rather how that RN captain succeeded. We kept on marching, the captain by our side twirling his gold-headed cane while the lynch mob dissolved about us. The Snowdrops had been conspicuous by their absence.

At last we reached the front and the captain steered us straight into the foyer of a hotel which served as a mess for US naval officers. There our captain called for their orderly officer, crisply gave him the facts and handed over the prisoner who was now protesting his innocence. We learned later that the entire matter had been a mistake. The coloured man had never been anywhere near any white girl. However at the time we were thanked by the RN captain with the following words,

"You may," he informed us, "cut along now."

Despite witnessing and even becoming involved in such scenes, we still found them unbelievable. Shootings at harbour side inns and near lynchings in the streets of a seaside town were not our kind of behaviour. In all fairness I should add that the vast majority of Americans behaved as well as they could in the unnatural conditions of those times. Many, like our US captain Mort, went beyond the bounds of friendliness. It was a very tense period.

This became increasingly apparent at our air/sea rescue base. Skip became more taciturn after each meeting with the other skippers. Timber alternated between going over every boat detail again and again or writing even longer letters to his wife in Northumberland. He had a third outlet. There was a seaman's mission hall along the quay with a snooker table in it. Timber liked a game and I would help give him one although I always lost. Fancy began washing and mending clothes as if his life depended on it. Paddy managed to obtain a supply of stout from somewhere and, against all orders, brought the bottles back to the boat or billet. There he would sit nursing a glass by the hour. Taff had chummed up with a pretty Welsh ATS girl from next door. It transpired that she came from the mining valley next to his. "Almost a foreigner, Glynis is," Taff told us. Sparks spent every spare moment with Sarah, slipping away with large loads of RAF material to re-wire her house. As for the Docs, they maintained their usual lubricities. In the case of Jock, our ex-forester, he began drawing trees. He was no artist, in that he tried to depict every leaf. At a later stage, Jock changed from trees to woodland creatures, but again he endeavoured to include every hair or feather.

The final days of May 1944 went by, then — from the first Friday in that fateful June — the weather rapidly deteriorated. June supposedly the best of British months, 'Flaming June' when holidaymakers used to sunbathe on south coast beaches, turned cold, wet and windy. The chill winds became howling gales whipping up the sea. Great waves battered shores and harbours where thousands of vessels were assembled. In coastal areas, sealed off from the rest of the country, the huge forces I have attempted to describe waited — impatient, anxious, fearful. Their world was restricted to a camp or airfield, warship or landing craft. They were severed from the normalities of everyday life, waiting to perform a stupendous task, perhaps to end their lives. On top of this came storms tearing away tents, tugging at tethered aircraft and battering ships against one another. As the weather depression deepened, so did morale of the combat troops.

We learned later of convoys — those that had to travel the longest distances — actually setting out in the appalling weather. They came down the Irish Sea and the Bristol Channel, rounded Land's End where conditions were horrendous, but pressing on as they had been ordered. They were then ordered to go back again. Others were diverted to harbours and inlets in the south-west. The problem was that every space where a boat could be moored had long since been occupied by other boats. Fortunately sailors are adept at helping one another, especially when an ill wind blows, and everyone did what they could. Personnel on the larger vessels fared best. Despite intolerable overcrowding, they were at least warm and dry. Those who really suffered were the troops crammed to capacity in open-topped landing craft. Many had been there for days and even longer nights. For them that first weekend in June was one of sheer misery. Down wind, their stench was indescribable. Aboard it must have been appalling. As well as seasickness and minimal sanitary conditions, they had to contend with sea spray and persistent rain soaking them to the skin.

During that weekend the weather conditions grew, if anything, worse. Gale warnings were issued and there was ten-tenths cloud overhead. Thus passage by sea and air seemed denied to us. No one had seen anything like it. The elements appeared to be giving us a warning. People talked of omens as thunder rumbled within the ink black clouds and lightning forked down to the hills. Aircraft were not flying, so gratefully we were not needed. We sat around our billet listening to the storm outside. By the evening of Sunday 4th June, the consensus of opinion was that the invasion would have to be postponed. That meant running down and overhauling the invasion machine before it was ready to operate again.

Yet unbeknown to us on that Sunday evening, the top decision was taken to proceed. After much checking and rechecking, the weathermen predicted a lull on the way. They called it a window in the weather. According to them, the wind would drop and the sky clear from Monday afternoon until mid-day Tuesday. Acting on such flimsy information, the commander-in-chief Eisenhower made his lonely decision. Tuesday 6th June would be D-Day. From that historic moment, another kind of storm, comprising orders and briefings, swept across England.

Our initial instructions came through at about 2 am on the Monday morning. We were to go aboard and prepare to put to sea. The convoys were re-assembling and would soon be on the move once more. They would require air cover from dawn onwards, therefore we were needed to provide air/sea rescue support. As we hurried through the old alleyways, Paddy remarked that there was no sign of the Grey Lady.

"No self-respecting woman," Fancy retorted, "would be out on a night like this."

For the gales still raged overhead. We could hear trees creaking above us and heavy waves crashing on the shore. Then we heard the roars of the duty and standby boat engines being started. They were gone before we arrived at the base. So too had most of the other ships from the harbour. We were told to warm up our engines then wait. An hour later, two more air/sea rescue launches put to sea and we left about thirty minutes after they did. By that time there was light in the sky and some on the water. A heavy swell ran green and cold under the cliffs of Portland. Beyond the Bill, we encountered high waves capped by thick foam. To us amateur sailors, the clouds looked as low and thick as ever, but Timber said they would lift. They did so grudgingly as we battered our way south eastwards across the Channel.

Admittedly we only played a minor part in the mammoth production named Overlord, yet the very nature of our role gave us a unique opportunity to see and observe. Once the decision to proceed was taken on the previous evening and orders had reached each unit involved, tremendous forces were unleashed. Thousands of vessels set out from every available point along the south coast despite the wind still blowing in fierce gusts and the sea continuing to run high. Over fifty great convoys, shepherded by destroyers and gun boats, were heading towards a zone below the Isle of Wight. This meeting place was officially called Area Z but became more popularly known as Piccadilly Circus. At first however, as we made for the assembly area, all we saw — in the words of the song 'Every girl likes a sailor' was the sea.

Suddenly we came upon a small landing craft lurching back towards England. Skip told Timber to circle it, then bring our boat as close as possible alongside.

"You seem," our skipper called to the landing craft, "to be heading in the wrong direction."

"Unfortunately yes," a miserable looking lieutenant replied. "We've taken too much water and our pumps have packed up. Not very seaworthy I'm afraid. Nearly sunk a couple of times."

We moved closer and could see dozens of soldiers baling for all they were worth with their steel helmets. They too looked miserable. It must have been like trying to exist in a half-full bucket of water.

"We'll keep an eye on you," said Skip.

Our launch accompanied the landing craft nearly to Poole Harbour before resuming course. Some time later we came upon a typical invasion convoy. Ahead of it were half a dozen minesweepers. They worked in diagonal formation, each vessel trailing a long, saw-toothed wire sweep to starboard. Theirs was a necessary but always dangerous job. If there were mines about, the sweepers would hook these out and destroy them — or be destroyed themselves. Behind the minesweepers, stretching back as far as the eye could see, came the convoy. The unwieldly landing craft slewed, rolled, pointed upwards and downwards while continuing to press forward. Some of the bigger ships flew barrage balloons and above these in the clearing sky we could see swarms of fighter planes in case the Germans tried to attack. It is amazing that, throughout the long daylight hours of Monday 5th June, thousands of ships crossed the English Channel without the Luftwaffe spotting them. Down below, everyone was on full alert, every eye watching and ear listening, every gun loaded and manned. Destroyers and gun boats raced about the flanks of the convoy, also ready and waiting for any emergency action.

During that day we were moved to a series of rendezvous positions off the western side of the convoy. While thus working our way south-eastwards, we saw more convoys. The normally empty sea was now seething with ships. There were mine sweepers, buoy layers, fuel tankers, tugs and coasters, hospital ships, cargo boats and marine transports, all heading in the same direction. Most numerous were the landing craft loaded to their gunwales with troops, tanks, trucks and guns. Every so often we sighted a great grey warship. Timber pointed out how they were flying their battle flags. The USS Augusta heavy cruiser, accompanied by the Texas, Arkansas and Nevada, led 20 American convoys. HMS Scylla, escorted by the Ajax, Enterprise and

Black Prince headed over 30 British and Canadian convoys.

Throughout Monday afternoon and evening the stupendous fleet made steady progress. As night fell the great river of ships divided into five streams, then into ten channels. There were to be two channels, catering for fast and slow traffic, aimed at the five beaches code named Utah and Omaha, Sword, Juno and Gold. Sparks learned through his wireless operator grapevine that allied air forces were making massive raids on the Pas de Calais area as if the invasion would be there and not Normandy.

By evening we were heading back to Weymouth to refuel and prepare for the next day's operations. The predicted window in the weather was opening if reluctantly. The storm clouds lifted a little higher and broke up over the sea. The Channel itself remained rough and we could not travel at any speed because of the rollers. At least we would be safely back in port by nightfall while the invasion fleet heaved to, waiting and wondering about the morrow. For a great many soldiers and sailors out there, this was to be the last night they would spend on Earth. It would also be an uncomfortable night.

We refuelled at Portland and went round to Weymouth Harbour. There we were told hot meals had been laid on for us at the house, but we must be back on board within two hours to receive special orders. I recall we were very quiet, hardly speaking to one another. On nearing our billet, Sparks mumbled something about having to see Sarah and, Taff disappeared next door for a few words with Glynis. The rest of us picked at the food provided. Many wrote letters which formed quite a pile on the hall table. All too soon the time came to return down the alleys to the harbour. While on the way, our ears picked up a faint yet powerful humming sound. We stopped to listen, could not make it out and moved on a few paces. By then we were high enough above Weymouth to see across the town and the now empty bay towards the Dorset cliffs and hills. Low clouds lay over the hilltops but, in between, the sunset had drawn a thin red line.

The humming sound increased to a continuous roar and we saw what looked like a star city approaching. The effect was created by the first wave of aircraft flying low over the hills. Each plane had on its navigation lights. Before the first line had set out across the sea, a second appeared, followed by a third and a fourth, fifth and sixth, wave after wave. These aircraft formations stretched over miles of sky, yet more kept coming. There were fighters alongside bombers, British machines among those of the Americans — Spitfires, Mustangs, Mosquitos, Thunderbolts, Typhoons, Lightnings, Halifaxes, Hudsons, Lincolns, Liberators, Lancasters and Fortresses — it was an aircraft spotter's dream come true. The noise from their thousands of powerful engines beat down on the darkening land and sea. The total sounds reverberated in the air, glancing off cliff faces and the sides of buildings, washing over our upturned faces, hurting our ears and even prickling our skins. We stood like statues, as did every single person that night in Weymouth, watching the huge armada pass overhead. Eventually Timber said,

"Come on lads. We'll be needed to support that little lot."

But it was not to be just like that. At the strangely deserted harbour, only air/sea rescue crews waited — some grouped on the quay, others in their boats. We talked to one another in low voices. The rumour was that we would be receiving sealed orders for D-Day operations. This proved correct. At last our skippers emerged from base headquarters holding buff envelopes. We were to put to sea at once, head across channel and open our envelopes at midnight. It sounds melodramatic now and certainly was then. The long awaited invasion of Europe was about to take place and we were involved. We were told to wear our life jackets and steel helmets. Jock had the first hour in the engine room, so I stood in the wheelhouse with Skip, Timber and the rest. We kept glancing at the envelope on the chart table.

Precisely at midnight, the skipper slipped a pencil into the top of the envelope, ripped open the flap and took out a single sheet of paper. We watched his face while he rapidly scanned what was written there, then Skip said,

"By dawn, we have to be in position five miles off the Cherbourg Peninsula."

There was not much to see outside. Occasionally a wave would gleam fitfully, its crest half luminous. Paddy produced tea. Taff sat on the wheelhouse floor, wedged in one corner. Sparks was as always entombed in his wireless cabin headphones on and listening intently. When I handed him a mug of tea, the sole response was an impatient wave of his hand. Down in the sick bay, Doc and Fancy dozed on two bunks amid an ominous selection of splints and bandages. After finishing my tea, I also dozed until Jock came and shook me awake. Then I went aft to the engine room, sat on the chair there and waited.

Came a pale, watery dawn and I was off watch once more. Timber had reduced speed because we began to pass through an increasing number of vessels. They loomed up to port and starboard, some long and low, some high and towering. One massive silhouette was a warship. It looked like a surfaced sea monster.

During the night a great deal had been happening. Decoy ships and aircraft again concentrated on the Calais area. The craft there towed barrage balloons to produce fleet-like signals on enemy radar, while the planes flew round and round dropping bundles of metalised paper strips thereby adding to the deception. Other ships and aircraft did the same near Boulogne. Yet more planes went up and down the centre of the Channel to bewilder the enemy further.

By that stage, allied pathfinders, parachutists and gliders full of combat troops began dropping on Normandy. We heard and read many of their tales later. What interested us more at the time was how our aircraft returned safely to England, took aboard as much ammunition and as many bombs as they could carry, before taking off once again for the invasion zone. They arrived on schedule as the warships commenced firing at the coastal defences. The combined onslaught was fearful. While naval shot and shell curved overhead, our planes flew miraculously through flak to bomb and strafe the same targets.

We had been told to expect heavy casualties among aircrews, but such was not the case. The main reason was the absence of German aircraft. Gone were the days when the Luftwaffe could hit everything and everyone with impunity. In fact the situation was now reversed. Yet what amazed us were the risks our airman took. Fighter planes went in at sea level. Light bombers kept passing over them. Then overhead and out of sight, droned heavy bombers unloading still more bombs.

So much was going on at the same time, it is difficult to put the complex happenings of that day into some sort of order. Individual pictures rather than sequences come to mind, though each had a purpose in the overall scheme. For example, submarines which had lain quietly underwater for days rose to the surface and erected masts topped by flashing buoys for the invasion ships to follow. Thus the troop-filled landing craft began making their way to the beaches. I can still hear those engines whining over the sea, also windlasses whirring as assault boats were lowered into the choppy waters. Loud-hailers blared orders.

"Get into line!"

"Keep moving!"

"Go, go, go!"

Many of the soldiers had to descend by wet nets and slippery ladders to the bucking craft. They moved slowly because of the weight each was obliged to carry. Apart from full battle kit, guns and ammunition, men had objects like hand grenades, stick bombs, radio sets and entrenching tools hung about them. The combined weight of body and equipment must have come to about twenty stone per man. No wonder they descended with care, particularly as ship and craft were being ground together by the relentless waves.

No sooner was an assault boat loaded with troops than it set off for the coast. Rows of pale expressionless faces passed us and we counted ourselves lucky not to be among them. The hundreds of boats on their way inshore left wide white wakes in the swirling green sea. By then more ships from over the horizon came to unload more troops. For once the sea did not appear empty and lonely. There were vessels wherever one looked — from majestic battlecruisers to fussy little tugs — all working with a single sense of purpose.

As D-Day wore on, the number of ships around us was progressively reduced. The first wave of troops had landed and was moving inland, apart from the American setback on 'Bloody Omaha'. The number of aircraft passing overhead also decreased. Occasionally we would see a squadron of RAF Beaufighters or US Air Force fighter/bombers keeping an eye on the area. We had neither received nor heard of any crash calls so, when the initial excitement began to die down, Paddy suggested making a meal.

"The guns, wireless cabin and engine room," said the skipper, "must be kept manned. Apart from that, we could do with some food inside us."

What happened was that Jock temporarily took Paddy's place at the aft Oerlikon cannon while soup and sandwiches were being prepared. After Jock had strapped himself to the gun, I talked to him from the nearby engine room hatch.

"We've seen history made today," I remarked.

"Och, if that's history," Jock retorted, "you can bluidy well keep it. Poor sods, blowing each other to bits. The sooner I'm back among my mountains and glens, the better."

Presently Paddy came aft with a tin tray on which were three mugs of hot soup and a plate of doorstep size corned beef sandwiches. Jock released himself from the cannon and we were about to sit on the after deck, when some more Beaufighters passed low over us. The following sequence was like a film in slow motion.

The leading aircraft banked gently and the rest followed. They made a wide graceful arc until all the planes were behind us, then the leader dived towards our launch. Suddenly we noticed bright flashes appearing in front of this aircraft and lines of little fountains rising from the sea just off our port quarter. Flashes came from the second Beaufighter before we realised what was happening.

Paddy was the first to react. Knocking the tray, soup mugs and sandwiches flying, he dived for the cannon. Jock and I likewise did some diving — for the engine room hatch. Once below, we both tried to get our backs against the single sheet of armour plating provided. At that moment, Timber rang down for us to start the engines. We were doing so when the Oerlikon cannon began firing on the deck above us. I had never been in the engine room during gunnery practice and was surprised at the pounding going on above our heads. I think our feelings were of complete disbelief. Surely we thought, RAF fighter planes would not attack a clearly marked RAF air/sea rescue boat? We were painted in traditional colours with large roundals on each bow, also flying an extra large ensign.

But such niceties were not for Paddy. He was being attacked, he had a 20 mm cannon at his disposal, therefore he would retaliate. Leaning back on the straps, swivelling the gun, he aimed at the next Beaufighter and opened fire. That was when Jock and I heard the pounding sounds above our heads. As the launch got underway, Timber spun the wheel, which threw Jock over one engine and me against the hatchway ladder. I decided to try to see what was happening and looked over the ledge. The view I had was of Paddy steadying himself for the next burst of fire. Behind him were three more Beaufighters coming in to attack us. As they did so, Paddy shouted,

"Come on you British bastards! One Irishman's worth more than the lot of you. I'll get you before you get me."

Fortunately the mistake was discovered at this stage. Suddenly the Beaufighters broke away and their leader began a slow fly past wagging his wings. But our Irishman's fighting spirit was up and the RAF plane was subjected to a prolonged burst of cannon fire. By that time, Timber had handed over the wheel to Fancy and was hurrying aft to sort out the matter. He shouted,

"Pack it in, Paddy! They must have thought us an E-boat."

"No one's going to attack me!" Paddy shouted back, still ready to fire. "Typical British. Always trying to wipe their feet on decent Irishmen. Well, I showed them."

"Yes you did, Paddy," Timber agreed. "We're shifting our position by a mile or so, in case there are any more mistakes."

"Let 'em try," said Paddy. "Just let them try."

We had barely settled down to more soup and sandwiches when a crash call came through. A Liberator had been badly hit over France, had tried to make it back over the Channel, but had ditched after fifteen miles north-east of our position. Everything else was forgotten as we reacted to the call. The course was set, the engines opened up and we went bouncing over the waves still rolling in from the south-west. After about twenty minutes, a second message came through. Another of our Weymouth boats had arrived at the scene of the crash and taken the airmen aboard. We were to resume our former position and await further orders. We did so in a more leisurely manner. The day was creeping by. On the invasion beaches, we knew, men were dead or dying. A large landing ship, heading back to England for a further load of men and machines, came our way.

"What was it like there?" Skip called.

"Like a bomb in a shithouse," came the Yankee reply.

We had no sooner taken up our former position when we were ordered to mid-Channel. As we headed north once more, the sea did its trick of suddenly becoming empty. We had left the sight of ships and the sound of guns for a watery waste. Green rollers from the Atlantic surged past us. It was a scene which summed up ninety-nine per cent of air/sea rescue work. We waited.

Late in the afternoon, someone suggested we pass the time by telling stories from our pasts. We had often done this, thereby revealing the most interesting and scandalous of subjects. Such sessions required a theme and it was decided the one for D-Day should be 'My most embarrassing experience'. Being the youngest member of the crew, I had to kick-off. I told an innocuous tale of how, when I was recently a schoolboy, I made my first date with a schoolgirl. An hour beforehand, I was bitten on the nose by a bee, but my friends insisted that I should keep the date despite my nose looking like a tomato. The crew chuckled good naturedly at this slight tale, which obviously would not be voted the best, before going on to relate their most embarrassing experiences.

Jock's story was diffidently told though it had more substance to it. Around where he lived, there was a stern minister who even equated whistling on Sunday with the flickering flames of hellfire. One weekday Jock had gone shooting rabbits. He saw, aimed and shot at a white object rising and falling in the grass. This object turned out to be the bare bottom of the minister. Luckily the lass below him was protected from the pepper shot, but it was most embarrassing, Jock concluded.

Next we were treated to the usual uxorious reminiscences from Doc, then Taff told of an urinating competition against a newly white-washed chapel wall. After that came Timber with whom it was difficult to connect a single embarrassing moment. However embarrassment, I discovered, extends over an extremely wide range — as the best tales told by Timber, Paddy, Fancy and Skip will testify.

"I went to sea as a raw lad," Timber began. "About the only job around Seahouses was fishing. As schoolboys we used to gaze out of the classroom window at the fishing boats arriving and departing. After lessons, we'd hang around the harbour, lending a hand where we could and trying to get into the good books of some skipper or other. North Sea fishing is a hard business and those skippers were hard men. There was one called Mr. Barnard, better known as the Barnacle. Most of the fishermen said he was the best captain there, so I resolved to be on his crew one day. It was arranged in those days by boys cadging sea trips at weekends and holidays. We were given the dirty jobs to do yet it was a good way to learn.

"Well, to cut a long story short, I was on one of these learning trips. We went down the coast to an area where there are shifting sandbanks. Mr. Barnard asked me if I had ever used a lead line and I replied "Yes sir". You always said "Yes sir" to the Barnacle. As you know, the bottom face of a lead line is hollow and filled with wax to pick up samples from the sea bed like sand or pebbles. I let down the line and, when it touched, called out the depth. But Mr. Barnard wanted to know what the sea bed consisted of, and I did not realise the lead could tell me this as well. So he came and pulled up the lead, turned it upside down, then asked, "What's lodged in that hollow, lad?"

"It looks like wax, sir," I replied. You can imagine how the others laughed. They still tell that tale in Seahouses. About me finding wax on the bed of the North Sea."

Paddy's embarrassing experience was a simple though unexpected one. The first time he saw a town was at the age of ten when his parents took him for a day in Cork. There his mother wished to look at the shops and his father wanted a jar or two. So Paddy was given money to go to the picture house. He arrived shortly before the performance commenced and, after purchasing his ticket, was led by a fair colleen into the auditorium. It was also the first he had ever entered. There were two aisles, he said, as in a church and practically every seat was occupied. Having nothing else to look at for the moment, the audience gazed upon the fair colleen leading the boy wearing his elder brother's cast-offs. The usherette reached the end of a row in which there was a spare seat and pointed it out to Paddy. Then, not quite sure of himself but being a good Catholic, our Irish lad genuflected and made the sign of the cross. A roar of laughter went up, Paddy concluded, making his ears burn long after the lights were lowered.

"My most embarrassing experience," Fancy began, "took place in the store where I worked. We floor walkers were obliged to wear formal clothes — black jacket and grey striped trousers. Every garment had to fit perfectly which meant tight rather than loose. We were inspected daily by the chief floor walker and woe betide anyone with unironed clothes or unpolished shoes. One pair of trousers I had were a little too tight, but we were rushed off our feet just before Easter so I did not have time to think about them. Suddenly Miss Moxam from Haberdashery hurried up to me, pulled me to one side and told me of a split in the seat. Immediately I backed against a counter." Here Fancy gave a shriek of recollection.

"I was horrified," he went on. "Fortunately Miss Moxam proved as ever resourceful". "There is no need to worry," she told me. "I always have needle and cotton handy for emergencies. The ladies staff room is empty. We'll go in there, you'll take off the split trousers and I shall mend them."

"Naturally I was rather reluctant about such an arrangement," Fancy gave another shriek, "though there was little else I could do in the circumstances. As you can imagine, I waited nervously inside the staff room door while Miss Moxam sewed up the split seam. She kept making somewhat saucy remarks which I ignored. To tell the truth, I was on tenterhooks. Well I mean to say — what if the chief floor walker learned of my being in the ladies staff room bare from the thighs down — apart from grey silk socks and chequered suspenders, black patent leather shoes and of course my spats.

"At that moment who should come heading towards the room but a bevy of female staff. I spotted them through a crack in the door which I held slightly open. Shutting it, I looked desperately round for somewhere to hide.

There seemed to be some cupboards at the other end of the room and I ran in that direction. Miss Moxam did ask what was the matter but there was no time to explain. I only managed to say "People coming" before opening one of the doors and backing through it. Unfortunately I backed trouserless into the Perfumery Department. The first I knew of my mistake was when screams went up from the lady shoppers." Fancy concluded his tale with a third shriek of his own.

Last but not least there was the skipper. In the past he had regaled us with stories of life at the top and this was to be a further example. To tell the truth, we were fascinated by such insights to the English class system. What else could Skip tell us to beat his previous accounts? He gave a slow smile and related his best tale yet.

"The worst required of us at Eton," Skip began, "was to have tea on a Sunday afternoon with our housemaster's wife. She used to invite four boys at a time and of course we could not refuse. Not only that, we were obliged to wear our best togs and highest stiff collars. To be invited meant a miserable weekend making all the necessary preparations. I should mention that the housemaster's wife was one of those women who appear too sweet to be true. She was nice to your face yet did not miss a thing. After visiting her, we were sent for by her husband who had been given a long list of our shortcomings. Like leaning back in a chair or reaching too quickly for the cake.

"The tea ceremony, as we called it, took place round a small oval Jacobean gateleg table designed for four. It meant that the six of us, including our housemaster, sat wedged close together. Too bloody close. Once seated, we could not move and indeed any fidgeting on our part would go on the list. The good lady served us from a side table sited slightly behind her. In order to turn towards it, she needed more space than the rest. This resulted in less space for whoever sat next to her, a position each of us tried hard to avoid. On the afternoon I am about to relate, yours truly was given the fatal chair.

"The housemaster's wife invariably managed to drag out proceedings for the best part of an hour. Her conversation was bright and her eyes brighter still. I picked up a plate of sandwiches to hand round and two fell off the side, which shortcoming I felt would be included in my report. As the hour dragged on, I desperately wanted to ease my limbs. Of course I could only do so from the waist downwards, such as crossing and uncrossing my legs. This represented a small degree of freedom which I exercised from time to time. While concentrating on what the lady was saying, I became aware of pins and needles, then one of my legs seemed to go dead. The trouble was I could not recollect whether my legs were crossed or uncrossed.

"I rubbed one knee and felt it respond. I tried the other but could not feel anything. A gentle massage of the kneecap followed by a squeeze or two produced similar negative results. It was the same higher up the thigh. I was becoming concerned when something took place to make me really worried. The housemaster's wife put her hand under the table, removed my hand from its exploration of her thigh and replaced it on my knee which was crossed over the other. All the while she chatted away as if nothing untoward was happening. She was also most charming to me when the tea party broke up. As you can imagine, I waited panic stricken for her husband to read out the report yet, this time, not a single criticism was made of me. Instead I was commended for my much improved behaviour. There must be a moral in that somewhere."

To me, those stories remain as much a part of D-Day as the spectacular events we had witnessed. Our session on embarrassing experiences helped to pass the long hours of waiting inevitable with air/sea rescue work. As we yarned, afternoon turned to evening and still we waited. We lay rolling on an empty sea under an equally empty sky. It was hard to recall that earlier in the day, the same sea and sky had been filled with ships and aircraft.

Eventually the return to base signal came through and we started our engines. We had taken part in the war's greatest exploit and now it seemed empty for us. D-Day however still had a surprise in store. While following a course for Portland Bill, we observed a lone, low, Flying Fortress, also on its way back to the Dorset coast. All four engines seemed to be running perfectly and the airframe appeared in pristine condition, the evening sun reflecting on the polished aluminium fuselage and plexiglass panelled gun turrets. Then, as we stood gazing upwards, what looked like a black dot shot out of the plane. A few seconds later we saw a parachute open, after which we could distinguish a man dangling below it. There followed one of the swiftest pickups in air/sea rescue history. We judged where the man would come down and, no sooner had he touched water when we hauled him aboard.

The airman was a tall, laconic Texan. He acted as if it was an everyday occurrence to jump out of a US Air Force Flying Fortress and be picked up by an RAF rescue launch. While stripping off, towelling himself down and putting on dry clothes handed him by Doc, the American treated us to an account of what happened — in a slow drawl.

"Ah'm a gunner, you see. We trained hard to become a great team, then what do those guys do to me? They leave me up there."

"Could you explain?" Skip prompted him.

"Sure. Suddenly they're not talking to me no more. Ah try talking to them but get no answer. So ah haul mahself out of the turret to find out what gives. Y'know what? Ah find the goddamed plane empty. They've gone, left me, vamoosed. Then ah really begin to worry. Y'know why?"

"You're not a pilot," said the skipper.

"Right in one! Ah can't land a Fort neither. When ah try sitting up front, the joystick seems kind of locked solid. How do ah unlock it? What do ah do next? Well, ah was figuring jumping out over land to be better than coming down in the sea, when ah see you guys. And here ah am."

We returned triumphantly to No. 40 Air/Sea Rescue base with our D-Day pickup — a life saved set against so many lost. As usual, we radioed ahead and the US Army Air Force had a car waiting on the dockside. The tall Texan was very grateful and worked our arms in turn, as if trying to pump oil. Perhaps he survived the war, returned home to Texas and became a multi-millionaire — in which case I hope he reads this.

Chapter 8

ENEMY ACTIONS

Over the next two months we were on operations seven days a week, twenty four hours a day. If that seems impossible, this is how it was done.

At four in the morning, the base duty watchman would cross the quay to our eight boats to give each a call. Timber was always the first up in our boat. He would keep banging with the flat of his hand on a bulkhead until everyone was awake and moving. Sometimes he physically shifted us, though the one person he never attempted to touch was our fighting Irishman. We were as reluctant to leave our sleeping bags as snails their shells. Washing in the two microscopic basins was not an overlong procedure. In the case of Jock and I, the engines required immediate starting and warming. The sooner we put to sea, the better the authorities liked it. In modern day parlance, we were pressured. So we shattered the early morning silence, especially round the town harbour. Meanwhile our skippers received their orders for the day at base headquarters before hurrying back to their respective boats. Then came the shouts,

"Cast off forward! Cast off aft!"

Another day's air/sea operation was about to begin. It was still high summer. At that hour light was beginning to wash over the Channel and gild the cliffs of Portland. There were we knew watchers on the heights, observation and gun posts, but from the sea nothing stirred. The only moving objects were eight rescue launches with white foam peeling away from their black hulls. As we neared the Bill, the boats fanned out, each heading for its appointed rendezvous.

By that time, our boat was pervaded by the aroma of coffee won from the Yanks, also Paddy's inevitable bacon fry. He and Timber were the only two to eat anything so early. The rest of us moved around like zombies, grunting rather than speaking. We had sorted out our ship positions long ago and automatically made for them — Skip on the bridge, Timber at the wheel, Fancy by the chart table and the rest of us in odd corners. We remained there as the eastern sky brightened and the sea changed its colour from grey to green. The Channel seemed to be empty. It was hard to imagine what we had heard taking place in France.

We had heard many stories. How out of all the American tanks that had come over the Dorset hills, only two survived. How after heavy initial casualties, our troops continued to pour ashore in their hundreds of thousands. How the Germans fought back despite the overwhelming aerial superiority of the Allies. These aircraft sorties brought us into the picture. We covered the English Channel with every available boat the RAF Air/Sea Rescue Service had available.

Once in position, we switched off engines and rolled as ever in the waves. That was when we had our official breakfast. Paddy usually added fried eggs to his bacon-sizzling activities and we would clamp the soot-flecked results between two hunks of bread. Afterwards we dozed, wedged in our accustomed places until the crash calls came. We tried extra hard for pick-ups, not only to save lives, but also because this meant the possibility of an earlier return to base. We had our successes, though most of those long days were spent cat-napping or keeping out of each other's way. The afternoons were noticeably longer than the mornings and the evenings longer still. When the return to base signal finally came, we acted automatically — starting the engines, steering and navigating as if in a trance.

We would return around midnight but, before getting our heads down, there was always much to be done. We had to refuel and that was an operation calling for concentration. It meant handling high octane petrol in hundreds of gallons. There could be no slip up, no spark from metal hitting stone, no naked lights from cookers or cigarettes. We helped one another manipulate and connect up the heavy hoses, check readings and clean up afterwards. By then the boats reeked of petrol which the short journey from Portland Dock to Weymouth Harbour did not dispel. Once moored up, we were supposed to be off duty. Theoretically.

In practice, there was much else to do. Fresh water tanks had to be topped up, rations brought aboard and, of course, work required on the engines to keep them going. We were also dirty and our chins decidedly bristly. So kettles were boiled aboard for some and others went ashore to the base washroom. Every two or three nights, a few of us would walk wearily up to the billet for a change of clothing after a quick bath or shower. People moved around in a daze. We were all too aware, as we carried out those duties and chores, how we must begin again in less than four hours time. First would come the early calls, followed by the same relentless round of operations.

Such was our daily life for those two months — out at dawn, back about midnight and off again four hours later. To start with, we had plenty of pick-ups, but the tempo slackened as the allied armies fought their way inland and air squadrons were transferred to France. Rescue operations eased off yet still remained arduous. By then our health was being monitored by RAF medical officers in conjunction with crew orderlies. As a matter of pride we tried not to report sick though some succumbed. On our boat, Taff was ordered to take a week's complete rest due to stomach cramps and Fancy developed boils on his bum.

At last it was announced that post-invasion activities were at an end. Instead there would be a duty boat with another standing by. Crews of off-duty boats could rest or go on leave. I was one of those who took leave and the train from Weymouth to Waterloo conveyed me to another deadlier world. It was the time when German V1 and V2 rockets had begun to fall on poor battered old London. At least one could see the V1's — or the doodlebugs as they were called — until their engines stopped. After that came the swift drive and sharp explosion. The V2's were infinitely worse. There was no warning. Life would be going on in some residential district until there came the sound of a heavy explosion. The brief period of silence which followed, was superseded by cries, screaming and dogs barking. After a couple of days and nights, I was glad to escape from London.

During the lull at Weymouth, we took the opportunity to overhaul our boats and engines. The former kept us afloat and the latter had run well, yet there was much requiring our attention. We did such work while not on duty or stand-by. Taff returned to us rather ashamed of his stomach cramps and Fancy of his boils. We began to lead a fairly normal existence including the resumption of some love lives.

The grand passion and main object of our interest was the Sparks and Sarah affair. He visited her at every available opportunity which meant most nights. Knowing that her husband was a captain in the commandos, we warned Sparks that the least he could expect was a broken neck. But Sparks merely smiled and said we did not understand. This seemed to be the case when, one Sunday afternoon, who should come strolling along the quay but the lovely Sarah and her tough-looking husband. He had boxer shoulders, a squashed nose and hands like mechanical shovels. The whole crew crowded on deck to witness the carnage.

"I have come to thank you," said the husband shaking Sparks' hand, "for your help at my house and for your friendship to Sarah."

"Your wife helped me when I was wounded," Sparks said, "so it was the least I could do." We looked on open-mouthed.

The skipper also acquired his latest spectacular during this period. He seemed to go for tall, slender women and the girl friend was a honey blonde American nursing sister. She was a female version of Jekyll and Hyde according to Doc who knew of her — stiff and prim on duty, loose-limbed and raunchy off. Among our crew she was remembered for two phrases uttered when Skip brought her along to see the boat. He came below first to make sure we were all decent, then brought his latest to meet us. While we gazed with ill-concealed envy at her shining blonde hair, sparkling blue eyes and ultra-slim figure, this all-American beauty looked round our cramped quarters then enquired,

"Jeez, how do you guys keep from screwing each other?"

I noticed that Skip winced a bit at this remark, but breeding told and he suggested I show the lady my engine room. She duly evinced interest in vee-drive gearboxes and dual circulation water systems. After what could be called a crouching rather than a guided tour. I went to the exit ladder and politely said,

"After you."

"No, after you," she smiled sweetly. "I've been down engine rooms before."

Despite their diverse upbringings, I believe the skipper was very struck on the American nurse. She made him laugh and they went around a great deal together. The rest of us managed at lower romantic levels. About this time, the ATS next door threw a party to celebrate our safe return and, a week or two later, we organised another to thank them. The grand piano was pulled from the corner of the big room and a Weymouth lady, who reputedly had been on the BBC, came to entertain us.

It is said that one's true character emerges with too much drink. Well our punch proved more potent than it appeared and affected me in an extraordinary way. When the BBC lady asked if anyone could sing, in order to accompany her in a duet, I claimed the honour.

"What can you sing?" she enquired after I had placed myself within the curve of the grand piano in a true professional pose.

"Anything," I replied.

We sang not one but several songs. It must have been awful for the audience, yet everyone was complimentary. The potent punch made up for the fact that I had never sung before, nor have I since.

We did drink a little too much in those days and it nearly had serious consequences. As I have mentioned, boat crews took it in turns for duty and stand-by. Those on duty remained aboard with the engines warmed at regular intervals. The stand-by crew was limited to the quay and its environs, which area encompassed base headquarters, the seaman's mission and three pubs. The pubs were most favoured and we would spend stand-by evenings over three, four or five pints. On the night when the following incident occurred, Timber was away having a well-deserved leave and Fancy had taken his place as coxswain. We were relaxed, happy and slightly oiled, when we heard the duty boat engines start up at ten minutes to ten.

The rule was that, immediately the duty boat left harbour, its stand-by must be ready to follow. Therefore we moved with alacrity. Our half empty glasses were left on the table. Outside, on the quay, one of the base staff was informing our skipper that a Mosquito fighter bomber with two airmen aboard had come down in Lyme Bay. We were to follow the duty boat because the bay was a big place to search, especially on a dark night. While I was starting the engines, Fancy kept flicking the signals to full speed ahead. The moment we were ready to move, he shot the launch out of Weymouth

Harbour at the rate of knots.

The sea under the lee of Portland Bill was like black glass. Our propellers worked with maximum efficiency in the still waters and the planing hull rose to glide effortlessly over the smooth surface. Conditions were both perfect and exhilerating. When I stuck my head out of the engine room hatch, I could hear singing up forward. "She'll be coming round the mountain when she comes". I had moved back into the engine room and was about to sit down when the boat seemed to take off. The bow rose sharply, the hull and propellers came out of the water causing the engines to race, then there was a fearsome crash.

Several sharp cracking sounds, akin to a machine gun being fired, took place around me. I managed to hang on, following which I shut down the engines. People were shouting,

"What's happening?"

"What was it?"

There is a tidal race off the end of Portland Bill that keeps altering its position. Sometimes the race moves close to the peninsula and at others it shifts far out to sea. Portland Race can be calm in rough weather and rough in calm weather. This is because the race does not depend on mere wind and wave. Instead it is created by powerful underwater forces, by deep currents flowing along Chesil Bank from the west to encounter those coming down the Channel from the east. Far below the surface these forces contend with each other. Later we were to examine the race more closely and comprehend what happened to us on the night in question.

Rising out of the seemingly placid sea were what looked like great ramps of water. Our launch had raced up one of the ramps at top speed, taken off and landed badly on another ramp. The impact of a boat hitting the ramp at top speed had parted some of the hull timbers. As for the sharp cracking sounds I had heard, they were our engines snapping their holding down bolts. But for the double plank construction, Portland Race would have claimed another victim.

Luckily I had held on to the engineer's chair which saved me from being thrown. The rest of the crew suffered nasty bumps and scrapes. While they were recovering up forward, I looked round the engine room to assess the damage. The starboard engine was resting on its bearers but had no bolts. The port engine was held only by one bolt. The centre engine still seemed secure. However the auxiliary engine, used for charging batteries and pumping water out of the bilges, had come completely adrift. Its exhaust pipe leading to outside the hull was ruptured in the middle. This meant we could not run the auxiliary or it would fill the place with dangerous fumes.

Then there were the batteries. They too had jumped and, while remaining connected, shattered the wooden boxes housing them. My main concern was the petrol system and luckily this appeared intact. I was checking the system pipes and joints when Skip and Jock came down the engine room ladder.

"What's the damage?" the skipper enquired and I told him. He then asked if we could continue to search for the Mosquito crew. It was a service tradition, also a matter of pride, never to abandon searches if one could possibly help it.

"I think we can go on," I said uncertainly. "We'll try the centre engine to start with, while Jock and I secure the other two."

"We're taking in water," Skip said next. "How about pumping it out?"

I told him it would have to be done by hand pump and that remains my main memory of what proved a long night. For the next six hours, we took turns to hand pump. Our search was in vain. Not a trace was found of the aircraft or its unfortunate crew. As we pumped, our thoughts went to the poor victims descending to that deep, dark and powerful current coursing off Chesil Bank. Throughout the night we were in touch with the other boat from Weymouth in case we had to be rescued ourselves. It was a humiliating situation and Fancy would keep on saying, "If only Timber had been at the wheel, it'd never have happened". I must confess the rest of us had similar thoughts, though Skip took an altogether different line.

"Portland Race is always changing its position," he maintained. "It was a risk we took — I took — trying to cut between it and the Bill. We wanted to reach the search area as soon as possible." Skip subsequently stuck to this account of the mishap and there were no comebacks.

In life however, one inevitably pays for mistakes some way or other. We limped back round Portland Bill next morning, with the duty boat escorting us, and were taken into the naval dockyard for necessary repairs. By that time, the US Navy had moved elsewhere and we were back to the Royal Navy with its own ways of handling such situations. As our boat was being hauled up their slipway, we came in for some scathing comments on our seamanship — or rather the lack of it. One observation was that the RAF would run into Portland Bill next.

We were obliged to take this and much more, as well as go through their system of long waits, paperwork and protocol. They did things their way which often bore no resemblance to the method we had been taught. For example, our three propeller shafts had to be checked for damage and alignment. An RAF engineering officer arrived, made many measurements and calculations, then told me to remove the propellers so as to inspect the condition of their keys and keyways. Normally each propeller was pulled off the tapered end on its shaft by an extractor tool. Two came quite easily but the third proved to be what was called a 'bastard.' The solution we had been taught and often used was to apply heat and penetrating oil.

When I mentioned this procedure to my opposite naval number, he reacted as if I was about to blow up the dockyard. "I would", he said, "have to obtain permission from his superior", and he doubted I would get it. He was right. There followed a series of interviews, climbing a kind of celestial ladder until I came before an archangel lieutenant commander. He acted as if my request was both incredible and mad. No, he certainly could not grant permission but he would take me to see his captain next day, for which interview I must wear my best blue.

The deity before whom I was ushered with due ceremony turned out to be a cheerful, red-faced gentleman who kept chuckling as the situation was explained. I wanted to chuckle with him, but everyone else remained motionless and expressionless. At last, after the matter had been well thrashed out, the captain made his pronouncement.

"We must help our RAF colleagues in their nautical endeavours," he said, "while adhering to naval rules and regulations. The relevant clause is concerned with fires and naked lights in His Majesty's Dockyards. As I interpret the clause, extinguishing possible conflagrations is the responsibility of the duty fire party. That indicates to me a condition of control." At the word 'control', people began nodding sagely which pleased the captain. He next asked me if the blow lamp I wanted to use was very big.

"No sir," I assured him, "it's very small."

"Then this is what I suggest. Our RAF colleague, with his very small blowlamp, should be surrounded by the duty fire party, fire extinguishers at the ready. When the blow lamp is lit, it must be kept under control by its operator and, as a second line of defence, by the duty fire party. It should also be put out the moment the necessary work is completed, otherwise the fire extinguishers will have to be used as per rules and regulations. The chief petty officer in charge of the fire party has to be briefed and I suggest Humphrey," this to the lietuenant commander, "you keep a discreet eye on the proceedings."

"Certainly sir. An excellent ruling if I may say so, sir."

"That's what I'm here for," beamed the cheery captain.

The exercise was laid on for three-thirty that afternoon which just about gave me time to make the Weymouth billet, put on my working clothes, then get back to His Majesty's Dockyard. There half a dozen ratings, holding huge fire extinguishers and supervised by a hulking CPO, stood waiting for me. Hovering in the background was Archangel Humphrey. When I lit the tiny blow lamp, a modern DIY enthusiast would have scorned, the duty fire party encircled me. Jock, who was helping, hoped the propeller would come away after all the fuss and luckily it did. Then, no sooner had the navy left, when Timber put in an appearance. Hearing about the mishap on leave, he had come straight back to Weymouth.

"What was that about?" Timber enquired, eyeing the departing sailors bearing fire extinguishers. I told him but he made no comment. Timber was more interested in the condition of the boat — or rather his boat — and he went over it inside and out. Fancy accompanied him full of apologies and was totally ignored. From Timber's point of view, anyone who damaged a good boat on a calm sea was an idiot. He did not say much, yet every word and gesture implied just that.

After we had finished work and returned to the billet, Timber asked me what I was doing that evening. I made non-committal noises and he suggested snooker at the seaman's mission. Fancy offered to keep the score but was still ignored. I wish he had come for I nearly beat Timber. Our first coxswain's mind was clearly not on the game. Towards the end of the evening he said of the boat,

"It's more like self-inflicted injury than enemy action."

By that stage of the war, enemy action had become remote. At last the Germans were being pushed back to their own country. True their garrisons still occupied the Channel Islands but these were watched from the French coast by the army and from the sea by the navy. The reasoning was why waste lives taking them when they would have to surrender sooner or later? We began to think nothing much more was going to happen, which is fatal in any war.

We were trying out our refurbished boat in Lyme Bay when an unexpected message reached us. It said proceed immediately to a position off the island of Jersey. Other air/sea rescue craft would be following us from the Weymouth and Lyme Regis bases. While Timber swung our launch in the general direction of the Channel Islands, Skip and Fancy consulted the charts. I watched them, anxious to learn what was happening.

"There!" said the skipper. He was pointing to a spot which looked perilously close to enemy-occupied territory. By then Taff and Paddy had joined us and the latter posed the vital question. "Why", Paddy wished to know, "would we be after going there at all?" "Sorry", Skip replied, "I should have explained" and he proceeded to do so. There were two destroyers operating out of Portland whose task it was to keep a stern naval eye on the Germans in the Channel Islands. We had often seen one or both of these destroyers at sea and they were thrilling to watch as they sped disdainfully past us.

"One of the destroyers," Skip explained, "has hit a magnetic mine and is badly damaged, while the other has been ordered to keep away. We are entering the minefield because we are the nearest, also made of wood."

"But our engines are metal," I said.

"Indeed they are," the skipper nodded. "We are about to put an interesting naval theory to the test."

I went to the engine room and relieved Jock. He pulled a face when I told him the news. Not only were we going close to the enemy-held Channel Islands, with their E-boats and shore batteries, we were also entering a mined area. Just when we had begun to believe the war was ending for us, it had returned with a vengeance. As yet we had no idea how the destroyer and its crew were affected. Mines were weapons we had seen or heard about. What they were like in practice was another matter. I ran through the engine room checks then stood half out of the hatch watching the sunset.

The sunset was a disturbed one and the sea in another of its moods. Heavy clouds were piled up in the west as if trying to smother the last light of day. Fiery hues glowed fitfully in the clouds and licked the waves. The waves themselves were not high. Instead they were curiously close together and marbled with foam. One imagined more deadly mines lurking below the disturbed surface.

Paddy came aft and strapped himself to the Oerlikon cannon. The gun had somehow become his property since he and Beaufighters exchanged shots on D-Day. We used to pull his leg about how the incident would become a legend in the west of Ireland. Now Paddy grinned at me, swung a full circle leaning on his straps and yelled,

"We'll be there in ten minutes. Hope I get a crack at them." He meant the enemy while I hoped he would not try shooting at the Royal Navy.

During those ten minutes the light faded rapidly. Suddenly Timber reduced speed and took the boat round in a wide sweep. I half expected Jock to join me in the engine room, but he was helping Doc prepare the medical equipment. So I had to concentrate on following signals from the wheelhouse, though I did manage to take an occasional look out of the hatch.

My first glimpse of the destroyer was towards its stern which was high in the water and at an odd angle. Soon we were passing by the bows or what was left of them. Armour plating and steel girders looked like crumpled sheets of paper and smashed up matchsticks. About fifty feet of the destroyer had disappeared and its superstructure lay twisted at all angles. As we drew alongside, rows of white faces appeared out of the gathering gloom. There was a slight bump, ropes took up the slack and we were by the stricken ship.

A silent second passed before pandemonium broke loose. I could hear hoarse shouts above the idling engines. After that came the thudding of boots on our deck. I estimated we were taking fifty or sixty aboard. In fact the total amounted to over twice the latter number, many of them badly wounded. As our launch was less than seventy foot long, the survivors occupied every inch of it above and below deck. By some miracle all were accommodated as Sparks called for another launch to come quickly and relieve us of the surplus.

I was still in the engine room, listening to the turmoil above my head and waiting for Timber's next signal, when a pair of bell-bottom trousers came throught the hatch. That first sailor was joined by at least fifteen others who kept giving me thumbs up signs as they arranged themselves in every available space around me. I could not help noticing how agile they were, also how the hot engine parts did not seem to bother them. They had come down because there was simply no room left on deck or anywhere else aboard. Our boat was now packed from stem to stern. Events were as usual turning out very differently to what I had imagined, yet the next event took me completely by surprise.

Still giving me the thumbs up signs, the matelots began producing various boxes and packets from under their clothes as if they were conjurers. One thrust a carton into my hand. It bore a label which read 'Nestles Milk Chocoloate Bars'. Another gave me cakes of soap, a third rolls of tobacco. While bemusedly accepting their gifts, my attention was drawn to frantic signals from the wheelhouse. We were to proceed at quarter speed. I made the necessary adjustments then looked again at my visitors. We could not speak because of the noise, but we grinned and patted each other on the shoulders.

In order to improve our communications, I took out the notepad and pencil Jock and I often used. The sailor nearest to me began nodding vigorously and indicated he wanted to write something. He scribbled a single word, which I could not quite see because the pad and pencil were passed on to the next man. I watched fascinated as the notepad went round the engineroom before being returned to me. Looking at the top sheet I saw that every matelot present had written the same word. There it was over and over again — in neat or scrawly longhand, in rough capital letters, all down the page and up the sides. It was the simple word 'Thanks'.

"Thanks for the presents," I turned over the page to write this. "Where did you get them?"

The sailor next to me glanced at what I had written, then crouched to use the engineer's seat as a firm working surface. Grasping the pencil tightly, he produced the following explanation in a careful hand.

"Always raid canteen before ship sinks. Old naval tradition."

This moved me. I felt a tremendous surge of admiration for these men crouching patiently and cheerfully around my hot, noisy engine room. They had lost their ship and personal possessions, yet they appreciated retaining the best gift of all, that of life.

I still had no idea what conditions were like elsewhere on board. The crowd of sailors around me, including two more on the ladder, precluded much movement. This state of affairs was suddenly to change. The wheelhouse rang down to idle engines, I heard another boat scrape alongside, followed by orders being shouted. Moving as one, the sailors scrabbled across the engines and shot up the ladder. I was left alone and, looking out, saw one of our Weymouth boats taking off those who were mobile. This was as well because, Timber told me later, we were dangerously top heavy with the weight of bodies. No sooner had the deck become clear when Jock joined me. He looked serious and remained so when I showed him our presents. Taking the pad, Jock turned over to a clean page and wrote,

"Bad up forward."

Presently I left Jock to see for myself. About a dozen sailors with cuts and bruises had found places to sit behind the bridge. Within the wheelhouse, the seriously wounded began. They were lying on every available piece of floor space and I had to step to each side of the blanket covered bodies. Some groaned whenever the boat shifted, but others lay ominously still. Below the sights were far worse. There was a sailor with most of his face missing, another drenched in blood and a young lad screaming continuously. Doc was doing all he could. He handed me what looked like a little tube of toothpaste, indicated the screaming lad and told me to give him a pain killing injection. When unscrewed, the top of the 'toothpaste' tube revealed a hypodermic needle.

"Push it well in," Doc advised. "Anywhere. Then squeeze."

I selected the lad's thigh. He took no notice of me when I spoke to him and pushed in the needle. The drug was morphine-based and acted quickly. The high screaming turned to low babbling, then there was blessed silence. Working my way round the bodies on the sick bay floor, I looked into the foc'sle where we usually lived. Every bunk including the letdown ones was occupied by wounded men. Most had now been sedated. An officer lay on the bunk in the skipper's cabin. His uniform and the skin below it was in tatters, yet strangely there was very little blood.

"I know I am going to die," he told me in a perfectly ordinary voice. "Too much gone. So don't worry about me. Look after the others. Take my blanket if you want." He had already pushed it off him to the cabin floor.

The officer was really dying — he went before we reached Portland — yet there he lay talking about it and being unselfish during the last moments of his life. If I had not heard him voice those words, I would not have thought such courage possible. Then I noticed Paddy standing beside me. Tears were running down our Irishman's face.

Such were the sombre scenes as we travelled back to Portland. Timber handled the boat with extra care so as not to discomfort the wounded. At long last we rounded the Bill and made for the naval base.

"I'll go to the floating dock," Timber informed Skip, "so we can just step off with the stretcher cases."

There followed an extraordinary episode that I can hardly credit to this day. We moored alongside the floating dock and prepared to take the wounded ashore. It was the best place, because the dock surface lay almost level with our deck. Moreover, a wide, slightly sloping walkway led to the dockyard proper. Paddy and Taff were moving the first stretcher case over the side when a naval officer came bustling down the walkway.

"You are not permitted to land here!" he shouted at them. "This floating deck is for use by Wrens only. Go over there." He pointed to another dock involving ladders and steep steps. "At once!"

This silly situation was rectified by Timber. Without a glance at the naval officer, who was rambling on about Wrens not seeing wounded men, Timber said,

"Come on lads. You know what you have to do."

After that, Skip took the naval type by the elbow and led him to one side "for a quiet word, old boy". By then ambulances had arrived and everyone, including Wrens, were lending a hand with the wounded, dying and dead. We were relieved when the last victim was ashore and we could cross the dock to refuel. Dawn was breaking by the time we reached Weymouth Harbour, but Timber had another task for us. He said,

"I know you're dropping on your feet. However this boat has to be cleaned before we go to the billet. No, we are not going to get someone else to do it. It's our job and the sooner we start the better."

It was amazing how quickly we cleaned up the mess. Torn clothing, soiled dressings and other unmentionable articles were put in bags then taken away to be burned. We washed the boat with carbolic soap and disinfectant, leaving all hatches and vents open for the air to blow through. In the end we were left looking at the pile of presents given in gratitude by the sailors. During the following days our skipper thanked us, the base commander thanked us and the navy thanked us. It was to be the most satisfying experience in our air/sea rescue career.

Life became less hectic for us as the main allied forces entered Germany and the enemy garrisons left stranded on the Channel Islands surrendered. Regarding the latter, it was agreed by radio for an unarmed German boat to meet an equally unarmed British boat at a specified position so as to initiate the surrender negotiations. Not only was the boat chosen stripped of all guns, security officers checked the entire vessel and the crew's personal belongings to ensure there were no weapons of any kind on board. When this boat reached the agreed position, it was to find the Germans fully armed. Their heavy cannon were manned, their sailors held rifles and their officers brandished revolvers. The Germans proceeded to search the British boat "to make certain you haf followed instructions", flinging everything about the place. After that they surrendered, which is perhaps a point the unilateral disarmers of today could ponder.

A few days later, our own boat and crew were selected to take four army staff officers, or 'brown types' in RAF jargon, to the French mainland. So light was the breeze and so calm the sea that the army men kept commenting on our enviable jobs.

"It must have been like pleasure cruising," one said.

When we reached the small French harbour, there were army staff cars waiting, also much snapping to attention and saluting from a squad of soldiers and two NCO's on the quayside. Once the staff officers were whirled away, everyone relaxed and we had a conducted tour of the half ruined farmhouse in which the soldiers were billeted. The district was devastated with a few French farmers trying to restore some semblance of normality. Rather than recross the Channel that evening, Skip decided we would spend the night in the little harbour.

After a meal, our skipper said he would take the opportunity to clear some paperwork while Timber wanted to write an extra long letter about visiting France. As others of our crew embarked on a marathon card game with the soldiers, I thought of going for a walk. An army NCO said it was safe if I walked eastwards, but anti-personnel mines to the west still had to be cleared. To my surprise Sparks, the one member of our crew who was seldom friendly and talkative, decided to come with me.

The wind was rising, which justified Skip's decision not to make the return crossing that night. It combed the dune grasses and sent sand whispering over our feet. Every so often we came suddenly on the debris of war — an armoured car in a ditch split open like an egg shell, a trail of ammunition boxes all empty, the half track of a tank revealed by traces of red rust in the yellow sand. Sparks began speaking about Sarah. How he had hoped to be back with her at Weymouth instead of wasting time in this forsaken place. How he would have slipped out of the billet and into her bed. How wonderful she was there. I muttered something about Sarah being married and Sparks rounded on me, his eyes blazing.

"What do you know about it? About him and her? About her and me? You haven't begun to understand people, let alone their relationships. Sarah's husband is a stupid boy who's never grown up. Schoolboy ways. Sports club type. He's a pig before he starts drinking and worse still when he's drunk. He comes from the right sort of family, went to the right school, knows the right people so they all say he's great — especially now he's playing at soldiers. Like that lot we've just landed. But he's not right for Sarah. I've knocked around and I know. What a sensitive woman like her must feel being married to him. He's a gorilla trying to play a violin if you get my meaning."

I got his meaning but, seeing no purpose in discussing Sarah, I asked Sparks what he did before joining the RAF. All we knew of him was he came from London. To my surprise Sparks began to talk freely of his background.

"We lived Lambeth way, as in the song, only it wasn't all fun and games. My old man had a barrow and, when he was short of the ready, which meant most of the time, mum went out charring. Anything to keep going. There were a lot of us kids. Also we had many uncles and aunts. Coming and going.

Rowing and making up. We made our own entertainment. A singsong. A knees-up. And there was our one week's holiday each year, fruit picking in Kent. The entire family, you understand, including old Gran. We moved like a tribe. Slept in barns or, if it was dry, outside. Ate what we could find or what the farmers gave us. Most of us kids were born nine months later, if you get my meaning."

"What did you do when you grew up?"

"Oh, buying and selling. Buy as cheaply as I could. Sell for as much as I could get. Strictly cash. We were all at it. You had to be real smart to win. Making a start was hardest but I was lucky. One of my uncles had a small yard and warehouse. I used to help him and, after his ticker stopped, I took over. Was a natural, if I say so myself. Within a year I was trading in the West End, then this war had to come along and spoil everything. Bombed my warehouse and home. Broke up the family. Made me a wireless operator on a bloody boat, I ask you."

We had walked some way from the harbour. A stream running into the sea forced us to turn inland where there was a network of fields, hedges and lanes. We decided to retrace our steps and took a path leading back to the coast. On our way we came upon a wooden hut with tar paper tacked over the windows and a piece of stick thrust through the door hasp. I was reluctant to go near the place, after some of the stories we had heard about German booby traps, but Sparks had no such qualms. He pulled out the stick, hinged back the hasp and pushed the door open with his foot. The hut had an earth floor and was piled high with old road signs. On looking closer at them, we saw they were German, obviously taken down after the invasion forces had passed. I saw no great significance in this, yet Sparks became strangely excited. He asked,

"Do you want to make some money?"

"Well, yes. . ."

"Right. We're both going back to the boat. I want you to get string or wire so we can tie the signs together in bundles. Leave the other part to me."

I did not know what the other part meant at first. Sparks went to see the soldiers in their farmhouse billet while I fetched a roll of locking wire and a pair of pliers. When we met again, Sparks was carrying some object in a canvas bag. On our return to the hut, he put the bag carefully to one side, then proceeded to lay the German signs between the dunes. At last all was revealed. Sparks opened the bag, took out a loaded Sten gun and raked the wooden boards with a stream of bullets.

Afterwards we wired up the splintered signs in bundles of ten and began carrying them back to the boat for storage in the after locker. No one took any notice of our coming and going. The card school was hard at it and Timber on the tenth page of his letter. Eventually the after locker was full and

Sparks suggested stowing the rest of the signs in the engine room. I had visions of the hot engines setting the splintered signs alight, but Sparks could be very persuasive. We did three more trips to and from the hut, at the end of which the engine room was almost as full as the after locker.

Next morning, when we started back across the Channel, Jock wanted to know what was going on. It was, I told him, one of Sparks' crazy ideas. Back at base there were certain difficulties to overcome — the main one being that our boat was moored right under the office and watchful eye of our commanding officer. We therefore had to move the signs that night. They first went to an RAF workshop under Weymouth Harbour bridge, then to a local crony of Sparks who handled transport for the navy. The signs, I later learned, were boxed as water pumps and travelled in official transport via Poole and Eastleigh to London. These various cronies and contacts of Sparks worked for hard cash and this is what I too received in about three weeks time. One evening Sparks took me to one side and handed me a wad of notes amounting to £64. It was, he said, my share of the German road sign business.

"What happened to them?" I had to know.

"Well they went to the Smoke," Sparks replied. "Vi my youngest sister works in one of them big offices the Yanks have near Whitehall. Full of Americans moving paper from one desk to another. So called servicemen who've never heard a shot fired in anger. No danger to their skins but no chance of glory either. Vi circulated a couple of sample signs and, before you can say "how's your father", the Yanks were lining up with wads of money. We did a roaring trade. Sold every sign. Even had a general in the catering corps buy two. Y'see he has a country as well as a town house. It was a perfect deal. The goods cost nothing yet sold at top prices. Next time I'll cut out the middle people like Vi and you."

The £64, a largish sum in those days, presented me with the pleasant problem of what to do with it. After due deliberation, I put £40 in the base bank, treated most of the crew to lobster suppers at Osmington and began buying a veritable library of Everyman classics at two shillings each, which books I have to this day. Sparks never would tell me how many signs there were, how much he charged for them or his incidental expenses. When I estimated over a hundred signs selling at £5 each — or was it £10, — he only smiled. But I do know that forever afterwards Sparks kept pestering the skipper for more trips to France. As he told me,

"I've got a ready market but no sodding supplies."

Over the years I have often thought about those German road signs. No doubt they were displayed in New York apartments, Boston mansions, Kansas farmhouses and Californian condominiums. One can visualise the centrepiece of a family living room, or perhaps Dad's den, bearing a Nazi notice with a bullet hole in it. And, when some admiring visitor commented on the sign, the proud owner would modestly talk about "enemy action".

Air/sea rescue launch 130 on patrol. Because it is travelling slowly, the craft is well down in the water and not up on its hard chine hull. (Courtesy R.A.F. Museum, Hendon).

Chapter 9

LONG RANGE

Foolishly we imagined the war was nearing its end and a grateful nation would be sending us home very soon. The Anglo-American armies were across the Rhine and the Russian steamroller coming the other way over Germany. Aircraft losses were low and we could not remember when we last went out on an air/sea rescue operation. So we kept our boats serviceable, as well as painted and polished, while awaiting the completion of our service careers. To become complacent is to provoke the unexpected. Just when our future seemed predictable, two changes occurred in the course of a single day. The first affected Sparks.

Our wireless operator had for some time been bracing himself for a showdown over Sarah. She had become more than another lady friend, but her commando husband was due to be demobilised like the rest of us. While Sarah represented the type of woman Sparks wished to retain, he did not relish messy divorce proceedings. At the start of the affair, Sparks often remarked how the mad captain was sure to stop a bullet or be blown up long before the war was over. Yet, despite highly dangerous raids and enemy actions, Sarah's husband seemed to bear a charmed life. He was one of the first British officers to reach the Rhine. There and then he decided to take a swim in that most German of rivers. While his men admiringly watched their crazy commander, he stripped off his clothes, dived in and was never seen again. Rhinemaidens or deep powerful currents must have pulled him ever deeper, then taken the body far out to sea. The news reached Sparks as we were about to hear of the other change affecting the whole base.

While Sparks was granted permission to go and see Sarah, we were summoned to a special meeting in the base office. There the CO told us that we would be going to the Far East for further air/sea rescue duties. Within two months, squadrons of long range boats with extra fuel tanks fitted were due to leave England from Plymouth. Meanwhile we had to attend courses at RAF Halton in Buckinghamshire.

The news caused consternation among certain members of our crew. Doc was indignant. How could Mrs Doc possibly follow him to the Far East? He applied for separation from our crew so as to remain in England but his request was promptly rejected. Sparks tried the same on compassionate grounds and, although the skipper backed his plea, there came another rejection. After that RAF lorries took us to Halton. We might as well have been chained in the back.

Halton was a camp noted for smartness and discipline. We were divided according to rank and trade, put into vast barrack blocks, then marched here, there and everywhere. The fact that we had been on almost continuous operations and built up tremendous team spirit did not impress Halton in the least. Jock and I had to attend a marine engineering refresher course. Long, long ago at Locking, our instructors had talked to us, even helped us. This time we had to sit through mornings and afternoons of training films. The darkened room and monotonous recorded voice kept sending one or other of us to sleep. Fancy, Taff and Paddy were given a crash course in cookery, the theory being we would have to fend for ourselves in the Far East. If there had been an emphasis on oriental food, then their course may have proved useful. Instead the lessons alternated between traditional English cooking and preparing tasty morsels.

"I'm telling you," Taff declared darkly one evening, "I'm never going to make any of them volley vents."

Luckily our stay at Halton was a short one. At the weekends I went for walks among the lovely Chiltern Hills and wished I had brought my bike with me to explore further. Then came our examinations with the dread of failing and having to remain in that establishment. However the RAF wanted us in the Far East, so we all passed and received our tropical kits. The crew became quiet and thoughtful from the skipper downwards. Skip had long lost his American nurse and fretted for her. Timber was concerned at missing out on the post-war revival of North Sea fishing. Fancy wanted to get back to his West End store, Sparks to Sarah and Taff to his Glynis. As for Paddy, he did not see why Irishmen had to keep on fighting for the British. Jock worried over his wife and bairns in the Highlands and the Docs were nearly frantic at the imminent termination of their conjugals. Secretly I was looking forward to the trip and stowed my bike in the after locker. This was just before we set sail for Plymouth to join the long range squadrons assembling there.

Our boat was allocated to the second squadron. It meant we watched the first boats leave, then heard they had reached Gibraltar and gone on to Malta. After that we waited a few days for the next spell of favourable weather. There was not much to see or do in Plymouth, for the city was still a mass of rubble from the bombing it received. After one depressing tour of inspection, we remained aboard gazing at the more inspiring sights of the Sound and Drake Island. At last there was a good forecast — wind light and sea slight — and we were off in line astern with pennants fluttering. The hills of Devon sank behind us as we approached Eddystone. Someone remarked we would be seeing much of the world before setting eyes on old England again. He could not have been more wrong.

It is approximately 1200 miles from Plymouth to Gibraltar, round Ushant and Finisterre, Capes St Vinçent and Trafalgar. We reckoned it would take up to two days sailing time, depending on the weather, and arranged a four hours on and four hours off watch system. We generally dozed during those off duty hours, slipping into our sleeping bags and raising the side pieces of our bunks. Safely contained, we could relax even when the boat met one of the bigger waves that had rolled right across the Atlantic. For food we had hot liquids and ship's biscuits. The engines ran perfectly so Jock and I could sit by the hatch watching the ever changing sea scenes. The waters became bluer with every passing mile. At night stars wheeled above our aerials and phosphorescent creatures moved through the deep. Even the most dispirited among us grew interested in such sights.

After the emptiness of Biscay, our landfall was Corunna. The coastal hills looked like those of Devon and Skip began reciting, "Not a sound was heard, not a funeral note. . ." He was obviously thinking about The Burial of Sir John Moore in the Peninsular War and I must confess that up to then I had no idea on the actual whereabouts of Corunna. This led to a study of the charts. One could almost drink the names — Torinana, Villa Garcia de Arosa, Viana do Castelo and of course Oporto. Sandy beaches shimmered below the hills. White houses with red roofs speckled the green slopes. Brightly painted fishing boats appeared and bounced around us, their crews waving.

Four hours on and four hours off. The Atlantic side of Iberia stretched ever onwards — Cape Mondego, Figueira da Faz, Santa Martinho, the Berlengas Islands. Skip remarked that his brother was at the British Embassy in Lisbon and it was a pity we could not visit him. Apart from having to remain in line, no service ships were allowed more than 24 hours in any Portuguese harbour. Although the European war was over, these neutrality rules had yet to be relaxed. So we passed the beaches of Estoril, the entrance to the Tagus, rounded Espichel and headed for Cape St Vincent.

The four hour spells went by swiftly, whether one was on or off duty. On duty there was much to observe and the sea fascinating at all times. Off duty we chatted or dozed. People came and went in a dream. Then it was time for the next shift. But first a glance at our latest position. We were at the base of Portugal, then a poor and desolate region called the Algarve. Who would want to live there? Land and property prices were at a minimum. During that watch, the pencil line on Skip's chart extended into the Gulf of Cadiz. Our skipper began quoting "Drake he's in his hammock. . ." and, as we neared Cape Trafalgar, it was Nelson's turn. "England expects. . ." A holiday spirit began to pervade the boat.

Spain and Africa were now closing in on us. Those coming off watch no longer wanted to doze. Yet another wonder awaited our eyes as we neared the Strait of Gibraltar. Dolphins in their hundreds appeared to escort our entry into the Mediterranean. They would turn slightly to regard us with friendly eyes, while smiling at our futile efforts to outrun them. For they were balanced on our bow waves as perfectly as birds on the wind. Occasionally one would dive to be replaced by another. To the left rose the bare hills of Spain, to the right dun-coloured African mountains and ahead, like a crouching lion, was Gibraltar.

We had made it without a single boat leaving the line, without one engine faltering. Now cruising speed was lowered to half speed, then quarter speed across Algeciras Bay. A trim naval launch escorted us to a dock. On stepping ashore to stretch our legs, we felt a glow of achievement. If the rest of our journey to the Far East proved the same, it would be the trip of a lifetime. But it was the August of 1945 and there came an inexplicable wait. Strange information began filtering down to us. A new type of bomb — something to do with atoms — had been used in Japan. It did not make sense at first. Only two such bombs were needed for the Japanese to surrender. Therefore both enemies, Germany and Japan, had been overcome. To all intents and purposes, the war was over.

013 moored off Rock of Gibraltar. This full side view shows wheelhouse with direction finding cone above and lookout crows nest on mast. *(W.D. Pereira collection)*

Winding up a global war however, poses more problems than starting one. Some days passed before we received news concerning ourselves. The first long range air/sea rescue squadron was being dispersed round the Mediterranean while we would be returning to England. No sooner had some of us become jubilant, this plan was modified. One boat had to remain at Gibraltar and our skipper pulled the short straw. We were shifted from the

(W.D. Pereira collection)

Long craft rescue craft powered by four Packard marine engines. This photograph was taken in Gibraltar VJ Day.

Crewmen grouped around Ace of Spades emblem adopted by RAF Craft 013. Flight Lieutenant L.R. Flower, later to become Group Captain commanding RAF Marine Branch bottom right. Author Wilf Pereira top right. *(W.D. Pereira collection)*

RAF marine base, where at least we had friends, to another dock opposite a still floating but dismasted ship-of-the-line. The great after cabin housed an admiral no less and our arrival opposite spoiled his view. We only had to appear with one sock lower than the other and a condemnatory note was brought over by a panting matelot. Fortunately our RAF superiors, mostly ex-pilots who appreciated air/sea rescue crews, were more tolerant.

They advised us to relax and enjoy ourselves. The war was over, the dangers passed and Gibraltar as good a place as any in which to find ourselves. We tended to agree for the first week. We could stroll round the Rock, walk to the top of the Rock or go through the Rock. The circular route was interesting and Catalan Bay at the other side most pleasant. The climb up the Rock was not arduous and the Barbary apes amusing. The interior of the Rock revealed a honeycomb of caverns with an extra deep passage, it was rumoured, leading under the Strait to Africa. Then there was the town's main street with excellent sherry at one penny a glass, flamenco dancing and a little concert hall where I heard a lovely young singer of Spanish songs by the name of Victoria de los Angeles.

At that time, before the diplomats sorted matters out, we could cross quite easily into Spain. Most servicemen visited the three places visible round the bay — La Linea, San Roque and Algeciras. La Linea was then full of bars and dancehalls, there was much poverty in San Roque and Algeciras was dominated by Spanish soldiery. One visit was enough for our crew members. They preferred to remain on board, swim twice a day, then enjoy a few penny glasses of sherry in the cool evening. It was at this stage that I remembered my bicycle in the after locker. I made enquiries. Not only could I take it with me into Spain, their buses would carry it on the roof for long journeys.

Such trips meant money at a level higher than that of service pay, but I still had most of my ill-gotten gains from the German signpost deal being looked after by the skipper. He handed over the cash and even toyed with the idea of accompanying me. However Skip had recently acquired yet another girl friend in the form of a raven-haired merchant's daughter who had vain hopes of making it to England. I went to her father's store and bought the civilian clothes necessary for my Spanish travels. The merchant, a swarthy Gibraltarian more British than the British, fitted me out with lightweight jacket and trousers, a change of shirts and socks, also stout shoes. After that I had to obtain the required papers. Came the morning when I approached a disapproving military policeman at our side of the frontier. He took a long time studying my papers, then remarked,

"At the best, you'll be flea-bitten from head to toe. At the worst, you'll end up in some stinking Spanish jail. If you do, don't expect us to bail you out." It was a surly send off.

The Spanish soldiers on the other side of the frontier were just as surly. So seemed the bus driver and passengers. I began to wonder if I had done the right thing. Fortunately, as the bus rattled on, people became increasingly friendly. They possessed the innate courtesy of their race. When food and drink appeared, I was asked to share with them. From that moment I knew my journey would be enjoyable.

Time has not dimmed the images of my travels in Andalucia. I remember dazzling white villages among the mountains, red roses and golden sunflowers, oleander bushes and eucalyptus trees, deer moving through cork forests and real eagles wheeling overhead. During the heat of the day, colours faded and there was almost total silence apart, from the occasional haunting sound of goat bells. Then, as the sun sank, the dome of sky turned deep blue and the rocks began to glow in rich tawny colours. Evening would find me searching for somewhere to stay the night, enquiring in tiled courtyards where fountains played. Supper was generally an omelette, coarse bread and red wine. Afterwards came magical night like black velvet obliterating the blemishes — the cracked walls, unpainted wood, pot holes in the roads. Inevitably a guitar would be played or gently strummed. I found the Andalucians friendly and honest. Their inn beds were hard but clean and, after a day in the open air, I soon fell asleep.

All however was not idyllic. It was then Franco's Spain and large obligatory photographs of El Caudillo hung everywhere. There were also menacing numbers of soldiers and civil guards about the place. Conversation ceased as they approached and only recommenced long after they had gone. The military, the clergy and the wealthy ruled the roost. Extreme wealth existed side by side with extreme poverty. I recall one opulent villa with beautiful colour tiles and wrought iron balconies. A wide verandah ran along one side for the rich family to take their ease — fat father, proud mother, pampered sons and daughters — enjoying iced drinks and glossy magazines. Below them, almost within touching distance, was a poor family in a home little more than a hovel. The man looked beaten, the woman ill and the children filthy. There was still a great deal wrong in that society.

Perhaps I formed the wrong impression. An English-speaking barman I met in Cadiz had certainly received a weird view of my country.

"How do you live there, senor?" he shuddered. "Once I had the offer of a voyage to England and accepted as I learned your language at school. The ship was going to obtain engine parts. It would stay one night during November. All that I understood. But I was not prepared for what I saw. As we approached the English coast, a thick dripping fog descended on us. Our ship, inside and out, ran with water. We had to reduce our speed to a crawl which made us late. When we arrived your bars were having their closing time. What a dreadful sight met our eyes. The houses were small and crammed together, the lights inside them dim and the pavements slippery from the freezing fog. We were glad to pick up our engine parts next morning and leave. I do not understand senor, how you English exist on such a hellish island."

"Where exactly," I enquired, "did you spend that November night?"

"Barrow in Furness," he replied.

Shortly after my first trip into Spain, I had a surprise at Gibraltar. The RAF kept a watchful eye on our health, which included dental check-ups. The surgery for this was in huts beneath the massive north face of the Rock. As I drew near them, a nurse opened a window, leaned out and waved wildly at me. It was Mrs Doc.

"I've made it," she said excitedly. "Flew in this morning and I am working already. There are so many nurses wanting to get home now the war's over, I had no trouble arranging a swop. I'm not really a dental nurse, but this will do for the time being. The other girls tell me we live in a hotel at the other end of the Rock. They say there's wonderful sun and sea bathing, the food's good and we each have a room of our own. Isn't it marvellous how everything has worked out? I gather your boat has been docked for weeks. That'll suit us, I must say."

Alas the Docs latest get together was due to be upset. Just when Mrs Doc had wrangled a Gibraltar posting and a room with a Mediterranean view, fate stepped in to take us away from the Rock on a series of navigational exercises. We had been grumbling at our static role opposite the admiral, so Skip decided to take action in his own way. He invited a superior naval officer from the admiral's staff to come and look around our boat, when the following conversation took place.

"I'm worried," our skipper murmured, "about morale. We were a good crew. Pulled well together on every sort of work — fighter support, bomber support, D-Day operations and, of course, there was that rescue from your mined destroyer."

"Mm," said the naval type.

"Lying here," Skip pressed on, "is sapping morale. What I have in mind are navigational exercises. Round and about Morocco, Algeria, Spain and Portugal. We could also do any little job the navy wanted, but the prime purpose would be to maintain crew spirit and ship efficiency. Most important don't you think?"

"Mm," said the silent service.

"There is another aspect," our skipper pointed out. "We seem to be stuck here, across the dock from your chief, upsetting him. He cannot help seeing us morning, noon and night. He compares us no doubt to some of the splendid ships he commanded in his time. Whatever we do, we're bound to make him bad tempered and that must rub off on his staff. So what better than to remove us from his sight?"

"Ah," said the naval officer.

Thus are great events decided. We were to visit Tangier, Rabat and Casablanca; Oran and Algiers; Malaga and Lisbon. Our non-air/sea rescue adventures would fill another book, but the first and the last will suffice. The crew was delighted to be on the move again — apart from Doc's "weeping and gnashing of teeth" as Paddy put it. Doc even tried to get out of our so-called training trips by claiming they were not true operations requiring a medical orderly. In reply to this argument, Skip simply said, "Be a good chap. Don't be tiresome" and that was that. So, at the earliest opportunity, we dressed ship for the benefit of the admiral and took ourselves off to Tangier.

As Tangier is visible across the strait from Gibraltar, travelling there did not present any navigational problems. By mid-morning we were moored beside the great mole made from huge rocks and stepped ashore for the first time on Africa. Tangier was then an international zone policed by a smart body of men to whom we reported immediately on arrival. They advised us to go about in pairs, to keep money in inside pockets and to hang cameras round our necks.

On our return to the boat, Skip said he would remain on board with Doc who, having Mrs. Doc in close proximity, did not need to see the fleshpots of Tangier. We then split into parties. Paddy and Taff wanted to find out what they drank in Tangier; Fancy, Jock and Sparks opted for the bazaars and, for some reason, Timber said he would look after me. I had survived my solo trips into Southern Spain but Timber still felt I should be protected from the dubious distractions Tangier had to offer. How a Northumbrian fisherman knew of these temptations was beyond me, but our first coxswain was quite clear on how to avoid them. Hanging his camera round his neck as advised by the international police, he said,

"We are going to stroll across town to that big hotel on the hillside, have the best meal they can provide, then sit in the garden enjoying a drink or two. Afterwards perhaps we'll look at some of their bazaars."

It sounded a bit tame beside the excitements such wilder spirits as Paddy had hinted at before departure. I was keen to leave right away, but Timber insisted on sprucing ourselves up for what he called a civilised excursion. At last, with our hair brushed, kit pressed and shoes well polished, we stepped ashore once more and set off along the quay. I sensed a dull time ahead with Timber yet our experiences were to be very different from those of the others.

Once off the quay and into the town, it was as if we had stepped back into Biblical days. Unlike the ubiquitous blue jeans of today, most of the inhabitants wore traditional Moorish clothes. The men were amiable and expressive, the women were not veiled but kept to themselves and the children were a delight. We saw our first camels outside a zoo and many magnificent horses. From an open space, which seemed to be a market place combined with a race track, tunnel-like streets led deeper into the town. I wanted to explore the bazaars, as the others were no doubt doing, but Timber was adamant. We were going to the hotel for a meal and a few quiet drinks.

Recently I saw the name of that hotel on a holiday brochure so I shall not reveal it now. Perhaps it is under new management or at least they have abandoned their post-war practices. Looking back on the episode and comparing it with my Spanish ventures, I can see the subtle difference. In Spain, I trusted everyone and came to no harm. In Tangier, Timber insisted on playing absolutely safe and came a cropper much to the crew's amusement.

The hotel manager was an oily individual with dark shadows under his eyes. He showed us to a terrace where there were several Tangier businessmen wearing western clothes. They sat in conspiratorial groups and conversed in sibilant whispers. The whispering stopped while we were being ushered to a table at the end of the terrace. This offered a fine view of Tangier including the harbour and our boat by the mole. The sight of the boat allowed Timber to relax a little. He and the waiter managed to communicate. We would have, Timber stressed, fried chicken and rice, followed by fruit we could peel ourselves.

After having and paying for this meal, Timber suggested we move to one of the garden tables. There he specifically asked for an unopened bottle of Scottish whisky with an equally new and sealed bottle of soda water. Timber then explained to me as follows.

"You remember I told you how I served my time with a Mr Barnard?"

"The Barnacle."

"Well he started on a Tyneside ship trading along this coast. The old Barbary Coast. They're still up to their tricks, he used to tell me. Have to watch 'em. When we were hanging around in the North Sea, he often spoke about the way of life over here. For example, fruit pickers use their left hands instead of toilet paper. That was why I insisted on fruit we could peel ourselves. Get it?"

"Got it, Timber."

"So you see, there's method in my madness. You may also have noticed how I carefully examined this bottle of whisky. Held it up to the light. Looked closely at the label and the cap. That was to guard against another of their tricks told me by Mr. Barnard. They have been known to empty a bottle of Scotch into four branded bottles, top up with wood or potato alcohol, then reseal them."

Timber liked a drop of Scotch. He could drink three or four doubles without them having the slightest effect on him. He used to do so slowly, savouring every sip. For my part, I preferred wines to spirits, but what could be more pleasant than sitting at our ease in a Moorish garden with the white town and the blue sea spread below us. Perhaps Timber had been right and chosen the best pastime.

Presently a strange sensation came over me. Instead of sitting in the hotel garden, I seemed to be floating above it, also twirling gently round and round. The feeling was in no way unpleasant. My mind was anaethetised to eliminate all fears. After revolving for some time, the dominant colours of the garden, houses and harbour began to blend — green and white, blue and green — before turning to dark grey. The fact that everything was getting duller, that night appeared to be arriving in mid-afternoon, still did not worry me. I could hear people's voices, the sounds of dishes and glasses. By way of conversation I told Timber I could not see.

"Neither can I," he replied, equably, "but there is no need to worry."

"I'm not worried."

"Good, all we have to do is walk downhill, through the town to the boat." He made it seem quite straightforward. "Come on them," said Timber taking hold of me, "we'll link arms."

In this ridiculous way, we left the garden seats and meandered down the sloping lawn to a door we had previously noticed in the lower end wall. We must have had a double squad of guardian angels watching over us that day. On reaching the wall, we felt along it until we found the door and were able to lift the latch. Outside lay a steeply sloping alley with a step every few feet. These steps were made of planks secured across the alleyway by pegs driven in the earth. Both the top edges of the planks and pegs protruded, yet never once did we trip. Unseeing, guided by our hands feeling for the planks or running along the side wall, we descended to a lower level without the slightest mishap. Someone spoke to Timber, though I did not catch the words, and he replied,

"We are all right, thank you. Perfectly all right."

Our self-confidence was so great that, at the foot of the alley steps, we stopped holding each other. Timber remarked we should not be seen returning to the boat arm in arm and I agreed. The worst was over, we were nearly there, best to proceed separately and casually as if nothing had happened. At this stage Timber's voice seemed to drift away from me and I could hear other people talking. In addition there were children shouting and the thunder of galloping horses. I decided to keep out of their way by making a slight detour. Alone and unafraid I began walking. The stroll took a pleasantly long time and at some stage I heard the sea. At another stage I lay down and fell asleep.

When I opened my eyes again, the night had passed and it was a bright, crystal clear dawn. The light poured past me, flowing over the sands and across the wide bay to distant Tangier. Strangely enough, my head did not ache or my eyes react against the already strong sun. I rolled over and looked the other way. There, standing motionless regarding me, was a man with three camels. It was then I thought of possible dangers and scrambled to my feet. While endeavouring to smile at the man who remained still, I checked to discover whether papers and other belongings were in my pockets. They were and, after an unreturned smile at the camel man, I began the long trudge back into town.

Tangier had started another day when I reached it. The market place was busy, children played and horses galloped. I made my way back to the harbour and along the mole. The skipper was raising the ensign when I arrived and he also elevated a quizzical eyebrow at the sight of me. I learned that Paddy and Taff had rolled in half an hour ago, while the police reported Timber sleeping peacefully in one of their cells. Over strong coffee we discussed our adventures of the previous night. Those I had shared with Timber were judged the most interesting and the crew, from the skipper down, could hardly await his return.

Another hour passed before Timber came sheepishly aboard. Apart from needing a shave, he looked normal and the camera still hung round his neck. We waited in anticipation while our first coxswain slowly shaved, washed and changed. Paddy handed him a mug of coffee which he drank in silence. After that, Timber announced he was going to the hotel where he intended to thrash the living daylights out of the manager. Skip laughed at this, adding that Timber would need a long arm as we were leaving right away. The journey back to Gibraltar was uneventful. We moored opposite the irascible admiral and resumed our in-harbour way of life. There was however a final sting in the tail for Timber. After using up the rest of his film on local scenes, he took the roll to be developed. When the prints came back, there were three shots of himself leaning bleary-eyed against a wall — wearing the full dress uniform of the Tangier International Police.

We went on several other much longer journeys, but they did not have the impact of Tangier. That and our final excursion to Lisbon were the real highlights. The 24-hour rule still stood, so we planned to dock about ten in the morning in order to leave by the same time next day. It meant travelling through the night, which proved a cool and pleasurable exercise, and we approached the coast amid swarms of fishing vessels. Once past ornate castles guarding the mouth of the Tagus, the river opened to a spacious lake with Lisbon set on the northern shore. Great squares led to even greater squares and little yellow trams ran everywhere.

The dockside was long and wide with more yellow trams — built Skip told us by the British — running from one end to the other. No sooner had a tram moved off than the next came into view, disgorging and taking on crowds of voluble passengers. There were further crowds round the boats unloading fish, also by the lorries and carts removing the catches — people talking, haggling and above all laughing. We were watching this animated scene when the skipper's brother put in an appearance. He arrived in a chauffeur-driven Rolls making its patrician way along the quay between the local inhabitants, yellow trams and baskets of fish. The limousine drew up beside our boat, the chauffeur opened a rear door and an older, larger, more expansive version of Skip stepped into view.

Whereas our skipper was well over six feet and weighed around fifteen stone, his brother would have tipped the scales at twenty. Skip's brother had a distinguished mane of hair surrounding an amiable mask of a face. He wore a cream coloured lightweight suit with a bright pink shirt, white collar and navy blue polka dot bow tie. Similar polka dot silk handkerchiefs drooped from his breast pocket and jacket sleeve.

"This," said the skipper as if apologising, "is Evelyn."

It was the first male Evelyn any of us had met, associating the name — in the words of Sparks — with some kind of old-fashioned bint. Yet this Evelyn possessed tremendous charm. Turning his expansive smile on our crew, he welcomed us to lovely Lisbon, told us he had heard so much about us he already knew us, then he invited us to dine with him that evening. When Timber ventured to mention guarding the boat, Evelyn airily replied he would lay on a couple of embassy chaps.

Thus for that one and only time we went out together as a complete crew. Timber reasserted himself by insisting we were spick and span for the occasion. He made Doc and Taff clean their shoes twice, also had Paddy's unruly locks cut by Fancy. Our skipper, whenever we encountered him in the confines of the boat, looked amused but he too spruced himself up. We were barely ready when two burly military types in civilian clothes arrived from the embassy to keep an eye on the boat. Then a squad car full of Lisbon police turned up followed by three city taxis. Skip's brother was proving most efficient as well as being a man not without influence in the Portuguese capital. We had been on a sight-seeing tour during the afternoon, now we wondered where Evelyn would take us that night.

That night must have cost either Evelyn or the British taxpayer a small fortune. First we went to a magnificent restaurant where our man in Lisbon was received like the King of England. After a superb meal, Skip's brother took us to a cabaret the acts of which went on and on, never repeating themselves, until the early hours. Finally Evelyn said he rather wanted a bit of a family chat with his brother but he would leave us in an interesting place within walking distance of the docks and boat. Again taxis whirled us through the city and brought our party to a large well-lit house. The double front doors were wide open and there was a feature which sticks in my memory. Two old fashioned footmen, wearing satin uniforms and powdered wigs, stood outside. Evelyn took us inside, settled us at a table and ordered a supply of drinks before departing with our skipper.

To begin with we were overawed by the opulent surroundings. Sparkling chandeliers hung from the ornate ceiling. Faceted lights shone from the mirrored walls. More footmen circulated among the wealthy looking men and beautiful women arrayed around the room. Champagne seemed to be the only way of quenching everyone's thirst with three buckets of the stuff arriving for us. The place looked like the foyer of a luxurious hotel — but without a reception desk. At the rear of the room were lift doors wrought from what resembled beaten gold. Two footmen stood on either side of these doors to usher people into the elevator. Suddenly Sparks remarked,

"I've cracked it."

"What do you mean?" one of us asked him.

"Have you noticed that while men and women keep going into that lift, only the women ever come back?"

At last we understood where the skipper's brother had brought us. We also realised that such an establishment was way beyond our modest means. Thus it was decided to finish the champagne and depart. However we had not reckoned on the waywardness of the Irish. Paddy informed us there was nothing like this in Ireland and, if only out of common courtesy, he would have to sample the local custom. As usual his logic escaped us yet, with Paddy, mere thought became instant action. He rose and made straight for a table round which sat several beauties. Dark eyes rolled and white teeth flashed as Paddy engaged them in earnest conversation. When he returned to us his mind was made up, though there remained one snag.

"It's a bit more than I have on me," Paddy explained. "However the difficulty need not arise if I you would all help me with a loan."

The loan took all the money we had but still came to less than half the sum Paddy had mentioned. When we pointed this out, he replied,

"Sure, but didn't we win the war?"

We watched fascinated, as did everyone around us, while Paddy sorted out the notes and stuffed them into his pockets. At this stage, Timber said a bit gruffly,

"We best be getting back to the boat, lads. Paddy, remember we sail at ten sharp. Be on board by then."

At the door we lingered to see who Paddy would select. He may only have had half the money, but with further Gaelic reasoning he decided on two girls. Paddy put his arms round their waists and the girls were laughing as they made their way to the golden lift doors. When the doors sprang open and the footmen ushered him inside, Paddy directed a wide smile in our direction. Then the doors closed and left us speculating.

We trudged back to the boat, weary after the many events of that memorable night. The police car still waited on the dockside and the two embassy guards were alert below. We thanked them, offered them refreshment but they had to leave. After they left, we took over the boat watch — first Fancy and Jock, then Taff and I. Taff kept wondering how our Irish rebel was faring while I nodded with weariness. Soon morning was with us, the docks came to life again, yet there was still no sign of Paddy. Skip arrived, we had breakfast and began preparing for departure. As the minutes went by, Timber became increasingly angry. By a quarter to ten, the Portuguese authorities were standing by to ensure we left at the allotted time.

The minutes ticked on, then the seconds. Skip and Timber discussed the situation. Should they ask for extra time so as to go and get Paddy? Would that blow up to a diplomatic incident involving our skipper's brother? Reluctantly Timber told me to start the engines. The noise drew a crowd who smiled and waved at us. We smiled and waved back, while trying to peer past them along the quay. The time had crept beyond ten and officials were tapping their watches. There seemed nothing else for it but to leave. Reluctantly we cast off forward and aft. Taff moved slowly, looking worried as he did so. He and Paddy had been good friends over the years, had been through much together. Was it all to end this way? The gap began widening between our boat and the quay when Skip called down from the bridge to Timber,

"Back alongside! Here he comes. . .I think."

There were people shouting and cheering on the quay as a car drove through them towards the boat. When the car drew near, we saw it contained a prostrate Paddy surrounded and ministered by about six girls. It was with some difficulty he extricated himself and, just as Timber managed to touch our bow against the dock, Paddy stepped aboard backwards, waving and blowing kisses at his recent companions. Then the boat was put into reverse and the gap again widened. Although Timber had endeavoured to create a smart impression of an air/sea rescue launch leaving harbour, the effect was spoilt by Paddy waving away, also shouting Portuguese catch phrases he had acquired. I was in the wheelhouse with Timber who told me what he intended to do with our Irish deckhand.

But events worked out quite differently. Paddy had made his last supreme effort waving goodbye to the girls. No sooner were we some distance from the dock, when he staggered to a sheltered piece of deck between the engine room hatch and our upturned dinghy. Paddy lay down, shut his eyes firmly and nothing anyone did could rouse him. Timber even used the toe of his boot, but Paddy merely moaned and turned over. In the end, Skip said we had better leave him alone, though Jock and I should keep a watch on the inert body. As we raced southwards and rounded Cape St Vincent, a freshening wind carried sea spray over the deck. Paddy took no notice of either the spray or of my tying an oil skin about him. Hours later, when we were entering the Strait of Gibraltar he rose, staggered down to his bunk and fell into an even deeper sleep. Many more hours went by before our Paddy surfaced. Recriminations followed on two levels — official and unofficial. Officially Timber wanted to put Paddy on a charge, which would lead to him being taken off our crew, but the skipper just said,

"Come now, Timber. Remember Tangier. I'll sort him out when he's fully recovered."

It was at this point Sparks initiated one of those elaborate service jokes to which we were prone. His contacts were on the admiral's ship as everywhere else. Paddy had regained consciousness and was sipping a mug of tea when two naval petty officers came aboard as if to see Sparks. During the course of their conversation, one of them enquired,

"By the way, who's the one to be flogged?"

"Ssh!" hissed Sparks, but they went on to talk about a wooden grid being set up on the old warship, of the cat-o-nine tails already soaking in a brine barrel and how a squad of marines would roll their kettle drums at each stroke. We thought Sparks and the navy overplayed the joke, but it registered with Paddy. After they had left he told us no matter how many strokes were given, the English would never be able to make an Irishman cry out.

It was at this stage that Skip intervened. He asked Paddy to take a turn with him along the dock. This they did, walking less than a hundred yards, in which short distance our skipper made his point. We never found out what was said yet Paddy became well nigh perfect afterwards.

By then we had been five months in and around Gibraltar. The English papers we received reported great changes taking place at home including a new government. Despite our so-called navigational exercises, we were becoming restive — even Doc, as one sort of exercise prevented the other. Timber wanted to get back to his fishing business and Fancy to his West End store. Then there were Jock, Taff and Sparks with their wife or girl friend interests. In the end we formally asked Timber to put our views to the skipper. Skip said he was agreeable to send a signal though he gave us due warning. The forces, or rather those who ran them, were notoriously difficult when it came to serviceman suggesting what should be done about their own lives. He advised against our request but, when the majority pressed for it, sent the signal. About forty eight hours later, a return signal instructed us to leave for Calshot and a home posting. Doc immediately tried to have Mrs Doc come with us. His request was turned down, which left her with the twin problems of how to drop the dental surgery job and to get home herself.

For some navigational reason it was decided we should depart at two in the morning. We spent the whole day getting ready, ignoring Doc's moans and preparing a surprise for the admiral opposite. To achieve it, we slackened the moorings which allowed more space between the boat and dock. Next we lowered the dinghy into this gap and took turns painting five foot high white letters on the grey harbour wall. Later, shortly after half past one next morning, we deliberately overprimed our engines so that, when switched on, the bangs must have roused the whole of Gibraltar let alone the admiral. We were a long way away by daybreak, idly wondering how the old salt would take our message. Fancy thought there would be squads of cursing matelots toiling through the heat of the day to obliterate the letters, while Skip pondered on the possibility of an RN submarine waiting to sink us without trace off Cape Trafalgar. For the three words we left behind us read KILROY WAS HERE.

Our journey home was copy book. We worked the usual four hours on and four hours off. The engines ran as smoothly as always, a tribute to their designers. Skip, Timber and Fancy plotted and pencilled our course round Spain and Portugal, then across a curiously placid Bay of Biscay to Ushant. There, off the western tip of Brittany, in an area notorious for foul weather, we witnessed a phenomenon which experienced sailors later told me only happens once in a lifetime. We encountered that rarest of conditions — a perfectly flat sea. Normally the bow would rise and fall on the slightest swell, but this ceased altogether during the night. The new sensation was akin to slithering over ice, even to flying through the air. It became more noticeable towards dawn when a light surface mist added to the illusion. The mist took the form of a thin diaphanous layer slightly above the water. Thus we could see below it, above it, also through it.

When the first rays of the rising sun reached us, the whole scene was transformed. The light was quite colourless to begin with, giving a pearl-like effect. Then a faint pink glow spread over the sky, through the mist and across the water. As the colour grew brighter and the magical effect more pronounced, Skip cut our revs then asked me to switch off all engines. Soon we lay still and silent. Our entire crew congregated on the foredeck marvelling. It was as if we were suspended in a pink glowing globe instead of an area noted for strong currents and ferocious storms. Not a ripple broke the glassy surface as the sun spread scarlet through the pink sky, sea and mist. I believe that was the moment in all those years when we came closest together as a crew.

The moment passed, like such moments do, all too quickly. The sun sucked up the mist and the red colouration was replaced by bright blues. We re-started our engines and headed towards the English Channel, old familiar territory. Soon we had entered another country of the sea. The green surge, the cotton wool clouds chasing each other across the Channel sky spoke to us of England being still there over the northern horizon. We visualised the coastline ahead — the hills of South Devon, the cliffs of Dorset, the forests and heathlands of Hampshire. They remained invisible to us as we were now approaching at a sharp angle from our last French bearing. Then the sea became choppier and spray began flickering on to the wheelhouse window. Suddenly Timber pointed. The rest of us peered forward but saw nothing at first. We looked again and at last noticed what looked like white spires protruding from the boisterous waters.

"The Needles!" Timber cried. "Surely you recognise them?"

Then we all did. We saw a verdant Isle of Wight hillside to the right and the low mainland coastline beyond. In half an hour's time we were running beside Calshot Spit taking in every detail. The camp and the marine craft section seemed strangely empty. We forgot the war had ended many months ago, also the last time we had seen Calshot was just prior to D-Day. The entire Solent had been filled with ships. Now the place was open, quiet, dreamlike.

Yet there were people about, old stagers who came to wave, take our mooring ropes and greet us. The RAF still functioned and retained control of our lives. Skip was asked to go and see the officer commanding, which took him away from us for over an hour. During that time we asked questions and were told of demobilisation taking place though very slowly. At last our skipper returned rubbing his chin, always an ominous sign. On stepping aboard he said,

"I did warn you about asking for a home posting. We've got one and they've told me where it is. It's to Stornaway in the Outer Hebrides."

(Photos by P. Dolphin from the collection of J. Pritchard)

Long range rescue craft looking forward. Note loud hailer centre, also port and starboard single Oerlikon 20mm cannons.

Long range rescue craft looking aft with twin Oerlikon 20mm cannons.

Chapter 10

HOME POSTING

The good news was that those of us who could make it were given a weekend's leave, also we could take the boat to Stornaway by any route. We discussed the matter and decided on the following course of action. Fancy and I would go to our London homes while Sparks and Taff visited their women in Weymouth. Skip still had his New Forest friends and Doc kept muttering about the return of Mrs. Doc. Meanwhile our reformed Paddy acted as boatwatch. At the end of these short leaves, we would journey to the Outer Hebrides via the East Coast and Caledonian Canal crossing Scotland. Such a route allowed Timber to see his wife at Seahouses, Northumberland, also Jock to visit his wife and wee bairns somewhere in the Scottish Highlands.

We had each acquired a few gifts from abroad — a bottle or two of sherry, wristwatches, perfume, leatherwork and jewellery. I took mine in a small holdall, giving the sherry to my father and visiting friends with presents. The weekend was soon over and Sunday night saw me — first on the train from Waterloo with Fancy, then on the Fawley Flyer — before walking down Calshot Spit around midnight. Our boat was moored well along the Solent but a duty dinghy ran us out to it. Ten minutes later I was asleep in my bunk. It was as if I had never been on leave. Others returned during the early hours but I did not hear them. We were still half asleep next morning when a heavy launch came alongside and someone bounded aboard. This someone wore oilskins despite the fine weather and had a little moustache that kept twitching.

"I am customs and excise," he announced. "I understand that this boat and crew have just returned from abroad. Therefore, before you go ashore, you are all legally obliged to show me everything you purchased so that I can assess if there is any duty payable." He sat down at the foc'sle table. "Now, each of you in turn will place your overseas purchases in front of me."

Timber kicked off, followed by Jock and Paddy. They showed the odd bottle, some wood carvings and children's toys. These looked insignificant on the table but the customs man carefully inspected every item before brushing it aside as being of little consequence. We were beginning to relax and Timber started to explain how some of us had already been home when our visitor went beserk. His moustache twitched all the more as he shouted,

"Home? You've been ashore? Without clearing customs? You've broken the law, do you realise that? Broken the law!"

Here our skipper tried to intervene, explaining how none of us had anything more than the inspector had seen. He apologised profusely, using all his charm, but the customs man continued to rant and rave. In particular he kept telling us about his powers of arrest. Fascinated, I watched Paddy's large hand rising to grasp the official warmly by the throat. It was just as well Timber gave our hot-headed Irishman a warning nudge. After that we stood around in an expressionless circle. Eventually the customs officer realised he was getting nowhere. He rose, pushed his way past us and stomped up on deck. Following him, we saw the customs launch alongside manned by a crew of three. One of them helped the official into its well deck and the matter might have rested there but for Sparks remarking,

"I thought Hitler had shot himself."

"I heard that!" the customs man started shouting once more. He jumped up and down, shaking his fists at us while the boat moved away from ours. "I can find out who you all are. Where you all live. I can search your homes. Tear them apart." The launch was some twenty yards away when the official directed his last and dreadful threat in our direction. Standing on tip-toe and shaking his clenched fists, he yelled over and over again, "I've got the power! I've got the power! I've got the power!"

We discussed him for a time, then moved on to a more interesting topic. Doc had informed us he was leaving the crew.

"I saw my doctor friend at station sickquarters. Spent a long time with him. He told me that, because the boat is not going on actual operations, there is no need for a medical orderly to remain on the strength. I gather there's a great deal to do here dealing with people being demobbed. So much so he's arranging for my wife to join me." Here Doc lowered his voice to signify that what came next was of great importance.

"A marvellous service is being evolved to look after the health of the nation. People will no longer pay for hospital beds and treatment, medicines and operations, glasses and dental care. Everything will be free. Absolutely free! Much however has to be done and who better to help organise the new service than my wife and me. After all, we are both in the medical profession and have its best interests at heart. Despite the war, as you know, we have always placed those interests before our own."

We managed to have a farewell drink with Doc, then put him out of our thoughts. The authorities were beginning to tread on our heels, wondering when we would take up our home posting at Stornaway. None of us had ever been there or had any idea what it would be like. Of course we grumbled about practically travelling to the Arctic Circle. Still grumbling we set off eastwards — first to Dover, then up the North Sea which proved as bleak as ever. Once underway, Timber kept the boat moving because he wished to reach Seahouses. Apart from wanting to see his wife, there was some business with his brother involving the purchase of a trawler. So we battled northwards against the metal-coloured waves.

Skip sent Timber ashore the moment we reached the little fishing port. We did not see our first coxswain that night or the next morning. At noon however, while we were refreshing ourselves at a harbour side inn, Timber walked in and said we were all invited to tea. The crew duly went, intrigued to see Mrs Timber the recipient of those long letters. We found her a cheerful woman, apple-cheeked and dimpled. The tea was a high one as they have in those parts, more than enough for a fleet rather than the crew of a launch, but we did our best. Later in the harbour side inn we met Timber's brother, also a wrinkled old man occupying the warmest place by the fire. Yes, it was Mr Barnard or the Barnacle.

"Honoured to meet you, sir," I said, unable to resist this opportunity of teasing Timber. "We've heard so much about you. How our first coxswain came to you as an apprentice. How he found wax on the bed of the North Sea. How you taught him about dealing with the North Africans."

"That's right," said Mr. Barnard in a crackly voice. "You've got to watch 'em."

"Like only drinking from an unopened bottle?" I suggested.

"That's right," he repeated. "Only from an unopened bottle. Follow that rule and nothing can go wrong. The tales I could tell you, lad. The tales. . ."

Next day, as we continued northwards past Holy Island to Scotland, Timber remarked how he had nearly taken me outside and thumped me for leading on the Barnacle. The others including Skip fell about laughing. We were acting more like civilian holidaymakers than a service crew. The war was fast receding into the past. Never again would we fish fighter pilots out of the English Channel, pluck bomber crews from the North Sea and take part in amphibious operations. We still resented the Stornaway posting as unfair, yet it was about to round off our respective RAF careers in a surprisingly different number of ways.

I suppose the first surprise to most of us was that a boat could cross the centre of Scotland, albeit diagonally. Shortly after leaving Inverness, we entered the Caledonian Canal with the Highlands, as it were, gliding past us. The scenery became increasingly wild and rugged, particularly after we came to the black primeval waters of Lock Ness. Once there, it was easy to imagine a monster lurking in those dark depths. As for the mountains and forests, it did not seem possible that anyone civilised could live among them. While we stood on deck, staring in awe, Jock quietly enquired if he could borrow my bike.

To this day I can hardly believe what we witnessed. Having got my bicycle and a few belongings together, Jock asked Skip to nose the boat into a certain spot along the north shore. There was no harbour, not even an inlet, but Jock seemed to know what he was doing. When we were in position, he went

Fairmile D-Class torpedo boat converted for RAF rescue duties. Note cut-out of starboard bow for previous torpedo launching. (W.D. Pereira)

Long range rescue craft passing through flight of locks at Fort Augustus on Caledonian Canal.
(W.D. Pereira)

down a rope ladder, until chest high in the icy water, after which the bike and bag were handed to him. Holding both above his head, Jock waded ashore.

On reaching the road, Jock shouted that he would rejoin us at Fort Augustus next morning. A side path we had not noticed before wound its way up a valley flanked by fir forests and craggy rocks. We watched fascinated as Jock pedalled and dwindled into the wilderness. Only after he was out of sight did we continue our journey along Loch Ness. The dark stretch of water with dark clouds pressing down on yet darker mountains made us wonder whoever could live thereabouts. The Scottish Highlands and Islands were to prove much stranger than we imagined. No sooner had we reached Fort Augustus at the western end of the loch when a monk arrived and said,

"The abbot invites you all to dinner."

013 in the Caledonian Canal making steady progress. *(W.D. Pereira)*

There was a monastery at Fort Augustus and the dinner they gave us proved more a medieval banquet. We had rich venison soup for starters, then lashings of roast beef and vegetables, followed by steam puddings with thick syrup. Strong ale was served throughout the meal at the end of which Skip somehow staggered to his feet for a thank you speech. In his humorous way he combined expressions of gratitude with references to holy poverty, fasting and abstinence, almost causing the jolly monks to roll off their benches. Afterwards we rolled back to the boat and miraculously reached our bunks.

Just as miraculously, Jock turned up early next morning. He was as fresh as a daisy while we kept groaning with our hangovers. In consequence we decided that Jock should operate the multiple lock system from Fort Augustus town — whose high street was the canal — down to the River Oich. We also began teasing him about the outlandish names in his heathen country. For his part, Jock merely grinned, made us black coffee and pointed out the sights we would otherwise have missed. There was Invergarry Castle, burnt by Butcher Cumberland after the Battle of Culloden. There was the Well of Seven Heads, where that number of murdered clansmen were washed in the local stream before being displayed. After Loch Oich, we descended through more lock gates to Loch Lochy, which we thought even more mysterious than Loch Ness. Its river, the Lochy, took us to Fort William via yet more locks known as Neptune's Staircase. Finally we came out by a sea lock.

It was just as well next day happened to be the Scottish Sabbath calling for a complete rest from our labours. It also happened to rain, a not infrequent occurrence we were told at Fort William. So we sat below listening to the rain, often augmented by sleet and hail, battering on deck. Towards evening the rain stopped, the clouds rolled back as in a theatre to reveal Ben Nevis. Its peak was capped with snow, which the setting sun changed from white to rose red. Red at night proved to be a shepherd's delight. The days that followed were, in Jock's language, "all bonnie."

On the Monday morning we set off at the same time as one of the local steamers but soon outdistanced it. For the next thirty or so miles, we travelled southwestwards along ever widening Loch Linnhe before turning into the Sound of Mull, The scenery on all sides was superb — towering mountains, glittering waterfalls and shining strands of sand. We put in at Tobermory on Mull, then rounded Ardnamurchan Point. Ahead, like emeralds in a sapphire sea, were the islands of Muck, Eigg and Rhum. They seemed to be floating and changing position as we made for the passage between Skye and the mainland. There were mountains loomed around us, more rivers and sea lochs became visible, more small villages and isolated houses. The names we read off the charts and quoted to tease Jock turned into poetry — the Sound of Sleat, the Kyle of Lochalsh, Scalpay and Raasay, Portree where we spent the night.

Then on the Tuesday morning we ran beside the wild coast of Skye until that magical island sank behind us and we were in the North Minch — a treacherous stretch of water as I was soon to discover. But that day it stayed calm and the forty mile crossing was an easy one. During the last part of our journey to the Outer Hebrides, the mountains of Harris — where the tweed comes from — rose along the western horizon.

At the same time, we came upon millions of sea birds. Some floated on the surface, swiftly ducking out of sight whenever they spotted a suitable meal moving below them. Others used dive bombing tactics, entering the sea like lances with hardly a splash. All of a sudden the birds took off, as if clearing a path for us. At first we could not understand the reason until the skipper from his vantage point on the bridge called us to join him. Heading straight towards us was a huge basking shark. Timber nudged the boat out of its way but the sea monster took no notice. After it had passed, the birds returned. The sea had shown us yet another of its different faces.

Stornaway proved to be a scrubbed little town lying well within a natural harbour. All eyes were on us when we arrived. Housewives came out of their homes and shopkeepers from their stores. Rows of fishermen leaning on harbour rails appraised our every movement. Later girls, in groups of two or three with linked arms, walked past us. Fancy and I, with Jock as interpreter, went ashore to purchase rations. While speculating on what these isolated Scottish shops would be like, the first one we came across was an Italian ice cream parlour. On impulse we went inside and ordered nut sundaes. We then sat at a window seat with a town's eye view of the harbour, our much travelled boat and the Minch behind it. Fancy said he had a feeling things would be different around these parts. He never spoke a truer word.

But first we learned the official reason for being in the Outer Hebrides. An American, whose rank we never did discover due to the part-uniform, part-civilian clothes he wore, came and briefed us. He had a prematurely bald head, thick horn rimmed glasses and was a fast talker even when smoking a cigar. The Sergeant Bilko television series could well have been built round our contact. We called him Fixit because he could fix everything.

"Whisky, you guys want some, I fix it. Harris tweed by the roll, I fix it too."

"About the reason for us being here?" Skip tried to pin down the American.

"Cover," said Fixit. "The United States Army Air Force is ferrying our boys back home via the northern route. From Prestwick to Gander, then on to the States. Every flight's gone right, but in case a plane comes down in the sea round these parts, we sure would appreciate rescue cover. Kind of insurance having you. Mind you, I reckon we'll never need it and you'll never leave harbour. In the meantime my dooty and pleasure is to look after you fellows. Show you a good time. Fix things for you. Say, have you ever been to a real Hebridean dance? There's one tonight on the other side of Lewis. I'll be along at eight with a truck. Gotta go now and fix it."

After Fixit had left us, Skip said that only half the crew could leave the boat in case we received a crash call. Those of us who did go to the dance were in for an experience. It began with Fixit arriving in a monstrous American truck with ample room for him and four of us in the cab let alone the back. Stornaway was not a big town and we were soon following a moorland road which seemed to be going nowhere. This road wandered up mountains and meandered along valleys, causing Fixit to keep changing gear. He also kept smoking a cigar and taking swigs from a flask.

Eventually we came to a high pass that gave us a westward view over the

Atlantic. The sun had long set but the sky still glowed and the surface of the sea was like burnished steel. A long narrow headland stuck out into ocean and at the end stood a large solitary hut. As Fixit steered the truck down the last slope, we noticed two or three old cars by the hut, also a couple of hundred bicycles. An eerie sound was coming from the hut, which prompted Fixit to make an odd remark,

"The Hebrideans are OK," he said, "in their peculiar ways."

Our arrival caused a lull in the proceedings. The band consisting of a drummer, fiddler and pianist were on drinking time. The people split into two groups — men along one side of the room and women opposite. There were many old folk among them and one crone rushed forward as soon as Fixit appeared. They embraced then he produced a bottle which she snatched from him and tucked away inside her voluminous garments. After that we were welcomed. Staring through us, the old lady said she could see we were good men, the savers rather than the takers of lives. We therefore had the honour of choosing the first dance. I remember Fixit grinning as he watched to see how we would handle the situation. For our part we selected every kind of dance we knew only to be met by total incomprehension. In the end our American friend said he would fix it and he did so by suggesting that the old girl should have the choice.

"Aiee!" she cried with delight. "The next dance will be a song."

Her song took all of twenty minutes. She wailed unaccompanied, though occasionally someone would join in or the fiddler would give her a few encouraging notes. After that the room came alive with people whirling, toes pointing and Hebridean eyes gleaming. The appearance of food did not interrupt proceedings. During all this, I noticed Fixit chatting with several characters. He also kept going into a certain side room. I sidled nearer to the door and saw him talking earnestly to a pretty lass. The situation however was not what I thought. It transpired that the girl acted as a go-between for what lay in a yet further room — roll upon roll of Harris tweed worth a great deal of money in those days of rationing. That was the reason why Fixit had brought the big truck. And us. We had to help load it then spent hours storing the rolls in various huts around Stornaway. Sparks tried to get into the act, but Fixit preferred the dimmer, more willing Paddy for further business dealings.

After the dance episode, I was determined to see as much as I could of the islands. Out came my bike when I discovered that buses would take it on their roof racks over the longer distances. My initial Hebridean ride was around Tarbert where Lewis and Harris are briefly joined. Afterwards I took buses and boats as far as South Uist and Barra, inhabited by Catholics unlike the Presbyterian north. Eventually I reached the end of the world. There was not a house to be seen, no one else within miles, just a concrete causeway barely raised beside the restless ocean. Occasionally a wave would splash across the causeway despite the sea being calm. What if one of these seventh waves Timber told us about came and sucked me out to sea? I was scared and turned back. Perhaps it was a premonition.

Back at Stornaway the weather continued fine and the nights mild. We had little to do as American flight operations slackened. Often we went on fishing trips using the 10 ft dinghy with its small outboard motor. One fine morning Skip thought he would take a trip down the coast and asked who wished to go with him. Fancy thought it would be interesting and so did I. We made sandwiches and a thermos of hot coffee, then set off. It was an idyllic outward journey. After a few miles we found an inlet with a solitary cottage at the end of it. Neither the crofter, his wife nor their two wide-eyed children could speak a word of English. Moreover they seemed distressed to see us. While Skip tried to explain we had come from Stornaway, the man kept waving us to go. In the end the crofter took our skipper by the arm, dragged him to the water's edge and pointed north-eastwards across the Minch. When he sucked his finger and held it aloft, we understood at last. The wind had changed.

We should not have attempted the return journey, but never imagined weather could alter so suddenly and drastically. Before we had gone a mile, the rising waves began slapping at our dinghy. It was all our outboard motor could do to keep us heading in the right direction. Skip tried steering us out to sea so as not to be driven under the cliffs. They had looked picturesque in the recent sunlight, but were now black and menacing below scurrying clouds. We huddled well down in the boat as spray increased over the gunwales. I remember Fancy asking the skipper if we were winning and, at that moment, the engine began to race. We knew what had happened. The pin securing the propeller to its engine driven shaft was sheared.

While the waves drove us nearer and nearer to the rocks, and as Fancy tried to row seawards, Skip and I hauled the engine inboard. What we had guessed proved correct. Not only had the pin sheared but the two parts were missing. Fortunately the propeller had not gone as well. "We must replace it somehow," said the skipper and — thinking on his feet — produced a silver plated propelling pencil we had seen so many times on the chart table. "Would this work," he asked, "if we hammered it in to shape?" Using one of the rowlocks, I hammered at the pencil without much hope. While I did so, Skip and Fancy started to bale out the dinghy, which was by then dangerously full of water. A silver plated pencil, I thought, will never hold where a steel pin has sheared. But it did. For mile after mile. We were also lucky in that the wind and waves did not escalate to a devastating Hebridean storm. As we neared Stornaway, the Eye Peninsula to the north began to protect us and, after passing the lighthouse, we could have been on a boating lake. Later Skip summed up this scary incident with his usual understatement.

"I might well frame that old pencil" he said "like one of Sparks' German roadsigns." He had known all the time.

It was around then we began to break up as a crew. Doc had already left us, but he had always been more Doc-minded than crew-minded. The next to leave was Paddy. He and Fixit became very close, always drinking and talking together. What they talked about eventually got back to us. Fixit we learned was planning to remain this side of the ocean. The British, he told Paddy, were pushovers when it came to doing business with a real hard-nosed Yankee. Paddy of course agreed and there was talk of profitable deals involving war surplus and scrap metal. To achieve this, Fixit needed a suitable partner and who better than our Paddy? It was Fixit who found out that, being an Irish citizen, Paddy could leave the RAF at any time. No time like the present, said Fixit, and so it happened. We had a farewell supper at a hotel and the following day Paddy departed on the ferry steamer. He was going to open Fixit's Manchester office. I hope he succeeded.

The next to go was Jock. He had seen the forestry people during that brief evening in the Highlands and they applied to the RAF for his release which was granted. I was particularly sorry to see Jock go, perhaps because I took him for granted. He had been on the boat when I joined it at Dover and remained a pillar of strength since. No task was too onerous, no job too difficult for him. While Jock was not the greatest conversationalist in the world, I had become accustomed to his quiet smile, especially when the going was tough. The going was not to be tough again, there were to be no more crash calls or long hours at sea, which was really why the authorities let him go. Before he went, we had another farewell meal at the hotel. Jock just sat and smiled throughout, with the rest of us chatting around him. Next day he too boarded the ferry steamer. All departures are sad and this one particularly moved me. I stood at the end of the quay for a long time until the ship, bearing my ever helpful colleague, rounded the lighthouse and disappeared from sight.

Thereafter the ferry steamer played an important part in our lives, including the mail that came and went on it. Everyone was either receiving or sending meaningful letters. Within minutes of the steamer docking, we would collect a bundle of mail held together by a special post office rubber band which snapped the moment removal was attempted. Invariably there were several letters for the skipper, ranging from official looking envelopes to dainty ones giving off faint but expensive scents. Timber also began getting large buff envelopes containing documents in connection with the purchase of the boat for himself and his brother. Sparks and Taff were likewise well and truly into the correspondence business. They read and replied to letters as if their lives depended on it. Sparks told me how Sarah was selling her Weymouth home because it contained mixed memories. With the money they planned to live in

Bournemouth where he would open an electrical repair business. In the case of Taff, our Welsh lad and his Glynis were equally determined not to return to the valleys — at least not yet. While we were at Stornaway he tried to join the merchant navy and learned they were laying off rather than taking on men. Taff then thought of staying in the RAF, as we were being bombarded by leaflets offering all sorts of incentives to do so.

One of these incentives was that suitable applicants would move up a rank. Taff a leading aircraftsman could therefore become a corporal. The irony of the situation lay in the fact that everyone aboard had been through years of war without promotion of any kind, let alone more money. No wonder servicemen used to sing "You'll get no promotion this side of the ocean." The whole system of jam tomorrow was in the highest traditions of British government and management.

Skip gladly recommended Taff for a position in the regular air force and, while the application was still being considered, we played another of those service jokes which might seem silly to civilians. To appreciate it, one must bear in mind our comrade was a typical Welshman, a member of a close-knit Welsh mining family, fond of music especially Welsh singing and now more than fond of a nice little Welsh girl. Fancy who masterminded the joke began by mentioning how Taff would have to take a postal examination. Sure enough a few days later, an official looking envelope, addressed to Taff, arrived in the bundle of mail Fancy fetched from the ferry steamer. In it was an even more official looking examination paper headed, 'All questions must be answered. Maximum time 2½ hours.' It also stated that the applicant should be supervised by crew members, so Taff sat at the foc'sle table with the rest of us around him.

"We can't give you a hand," said Fancy, "but read out the questions."

"Question 1," Taff read slowly, his eyes widening. "Write an essay on family life in a Welsh valley. Question 2. Describe a typical shift worked by a Welsh miner. Question 3. List ten of your favourite Welsh songs. Question 4. Give your real reason for wanting to marry a Welsh girl." At this point, Taff stopped reading then yelled, "Bastards! You rotten bastards! You've been having me on."

Not too long afterwards, the tables were turned on Fancy. He had managed to get through his service career without too serious an involvement with women. Because of his matinee idol looks however, many a girl tried to become involved with him causing Fancy to keep up his guard. Our second coxswain must have lowered his guard on reaching Stornaway, for that was where romance finally struck him.

There was a certain shop he often visited owned by a dour Hebridean. This shopkeeper was an elder of the church with an equally dour wife. They in turn had a 25-year-old red-headed daughter who lived, worked and worshipped with them. The first we knew of the liaison was when Fancy mentioned he had been invited to tea. High tea. We began pulling his leg and the knowledgeable Sparks informed him that the hottest women in the world were red-headed Scottish Presbyterians. This made Fancy very angry, so we knew it was serious.

About three weeks later, Fancy said he was again going to tea with the daughter who we now knew as Agnes. Apparently her parents were spending the day at distant Tarbert, so she and Fancy would have the house to themselves. This time Sparks refrained from crude suggestions, but looked worried. We sat around watching as Fancy pressed his best blue uniform, then dressed with extra care. He left us on a quiet afternoon and evening. Skip went to the hotel to make an arrangement which will be featured next; Timber sat frowning over more papers from solicitors, while Sparks and Taff were intent on replying to letters sent them by Sarah and Glynis respectively. I lay reading on a sick bay bunk and was therefore the first to see Fancy when he fell, rather than descended, the companionway steps. His hair was dishevelled, there were red smudges over his face, scratches on his neck and his recently pressed best blue was in considerable disarray.

It took a stiff whisky to return Fancy to the borders of coherency and, even then, his story was disjointed. Sitting in the sick bay with Timber on one side of him and Sparks on the other, he gabbled about traumatic experiences.

"Her parents were not due back until the last bus, she kept telling me. There was a fire burning in the grate and the sofa pulled near to it. Agnes had made the tea herself. Warm scones with cream and two kinds of jam. Also two kinds of cake. Afterwards I helped her wash up then, I thought, we might have a little cuddle. It was as if I had touched a switch. The next moment she was all over me. Taking off my tie, undoing my shirt and trousers. Hands everywhere. Clothes everywhere. She was stronger than me as well. Never known anything like it." Here Fancy uttered what we came to regard as his most memorable expression. With eyes staring, he declared,

"She pinned me screaming to the sofa."

By that time, Skip had returned aboard. He stood quietly beside us while Fancy continued with the lurid details. Apparently it had been two hours before Agnes allowed our second coxswain to leave. It was like hearing about the mating habits of a female tarantula. In the end our skipper insisted on giving Fancy his cabin for the night, together with a hot drink and a couple of sleeping tablets.

I was on deck next morning when who should come marching along the pier but Agnes. She was formally dressed in coat, gloves and big handbag, also wore a hat with a little veil on it. Agnes informed me she wished to speak to "her Frederick" and for a moment I forgot that was Fancy's Christian name. He came on deck readily enough and walked with Agnes to the end of the pier. They remained there for a long time with us watching from the wheelhouse.

Fancy stood at the very end of the quay with Agnes positioned to cut off his retreat. Between the she-devil and the deep blue sea, was how Sparks put it. Then unbelievably they came past us and went to buy an engagement ring. At the modest party thrown later by her parents, Agnes informed me how she now had the key to Fancy's London flat. It was behind Baker Street, she said, and she was soon off to see what could be made of it. As Agnes had never been to the mainland in her 25 years, that represented a fair challenge.

We were still bemused by these happenings when our skipper had his lady visitor. The lovely Fiona of the Scottish castle came over on the ferry. We stood, stared and marvelled at the slim figure waving from the steamer. Fiona wore a dark green velvet suit with a matching cap bearing a silver thistle and feather. On arriving down the gangway, she and Skip briefly embraced before hurrying off to the hotel. Not long afterwards our skipper returned to say he would be out for dinner. To our surprise he came back early and retired to his cabin. It was most intriguing.

Next day, about mid-morning, Fiona paid us a visit. She wore a dark blue outfit this time, which did even more for her golden hair. Yet we could not help noticing how the lady seemed subdued, as indeed was our skipper. We were invited to the hotel for lunchtime drinks, then left them to their meal. In the final hour and a half before the ferry steamer departed, Skip and Fiona went for a stroll along the sea shore. We could see them walking well apart kicking at pebbles. Finally they went to the hotel, collected Fiona's bag and she boarded the steamer.

Skip was not the sort of man to encourage or answer personal questions, so we just had to wait. While waiting, our visit to the Outer Hebrides came to an end. We were to proceed to RAF Calshot where our boat would be decommissioned. The crew, with the exception of Taff, was due to be demobilised. As we were much depleted in numbers, it was permissable to make our way southwards by easy stages. At first we thought of returning through the Caledonian Canal and along the English East Coast so that Timber could again visit his home town. I think it was the thought of all those lock gates that deterred us. Instead it was decided to take the western route between the Scottish Isles and via the Irish Sea so we could re-visit Weymouth for the benefit of Sparks and Taff.

We travelled southwards at a leisurely pace. It was more a pleasure cruise than a service exercise. Everywhere we witnessed moves being made towards peacetime activities. There were still shortages, rationing and austerity, yet most steps were in the right direction. As for the sea on that trip, it behaved like a child showing how good it could be. The winds were mere zephyrs and the waves no more than playful. Even the waters off Lands End were calm with just a token swell.

Weymouth we found was bustling and anxious to resume its family holiday status. Wherever one glanced, repairs, repainting and refurbishments were taking place. Some of us strolled to our old billet but it was a different road, occupied by families with children playing in the gardens. We began to feel like anachronisms in our uniforms. As for Sarah and Glynis, we met each of them that day. Although still extremely attractive, Sarah looked much older. She was staying in a small hotel waiting for Sparks and their move to Bournemouth — her money at the ready in a nearby bank. Sometimes I think women trust men too much. In the case of Glynis, civilian life had transformed her. The ATS khaki uniform used to do nothing for her dark hair and sallow complexion. Now, with a little make-up and a primrose yellow dress, she looked as pretty as a picture. Glynis was working in a local chemist and loving it. She was certainly proving to be a capable girl, well able to help Taff in his RAF career.

Later I went walking by myself along the front and round the bay to the cliffs beyond. From there, one had the best view of Weymouth and Portland. I tried to visualise the bay below me filled with warships and landing craft, the hills with armies and the sky full of aircraft. Somehow I could not do so. It was as if nothing had ever happened in that quiet countryside — apart from a few minor disruptions from Romans, Normans, the Napoleonic Wars and the D-Day invasion. Night was well advanced as I returned slowly by the murmuring shore.

Next day we did our last journey together back to Calshot. Timber took us at half speed as if reluctant to reach our destination. We found the marine craft section more deserted than ever, apart from the lines of redundant boats swinging on their Solent buoys. We too were given a buoy and told to stay aboard until our vessel was de-commissioned. This proved a dispiriting process. No one's heart was in it. Timber, Sparks and I had to plough through long lists, page after page of items, with civilian clerks who were as bored as we were. There were discrepancies, losses and arguments that Skip smoothed over in his usual mild manner. At last came the signing session which meant the old boat was no longer ours. Next morning we would go ashore and wait at the upper camp until we were sent elsewhere or demobilised.

Sitting on deck, I again found it difficult to visualise all which had happened aboard. The crash calls, the long searches, the even longer hours at sea, it was as if they had never been. I was still sitting there when Skip came and sat beside me. He asked me what I intended doing in civilian life and I talked vaguely about training for something.

"I'd really like to be a writer," I concluded, "but I suppose one must first have a steady trade or profession."

"Yes, it might be the best way," Skip nodded. "Keep writing in mind however. I've noticed you watching and listening. If ever you write, do so from recollection. Trust more to imagination than to memory."

"That's rather good!" I said.

"Not me," he smiled. "Samuel Taylor Coleridge."

Skip then started telling me about his future. His brother had opted for the diplomatic service so our skipper felt he must shoulder the family responsibilities. I found his usual understatements hard to follow. "Father was definitely eccentric. Mother did her best. Bit of a muddler though. Despair of solicitors. Fortunately a few good managers kept things going. But the war didn't help. Will need time to sort out. Have to stick at it. Get married. Carry on the old line. That sort of thing, y'know". I nodded wisely, without having the slightest idea what Skip was trying to tell me.

Next morning we left the boat and went to the upper camp orderly room. Inevitably Skip and Timber landed orderly officer and orderly sergeant duties while Fancy and I were put in charge of two huts unexpectedly filled with West Indians. Taff was due to take a corporal's course and, though we did not hear it at the time, they told Sparks he could leave the air force when he liked. He immediately cleared camp without a word to us.

There was a letter awaiting Fancy, written by Agnes from his London flat. He read it to me, about the place being too small for the two of them. Fortunately property was cheap and available at the moment, so she was looking round on their behalf. It would be a good idea to have a joint account. She had had words with a Presbyterian minister.

"I so much admire her," Fancy informed me.

"You do?"

"Why yes," he seemed surprised. "Think of the giant step she has taken. From the Outer Hebrides to the West End of London. There's character for you. Agnes is what I need. A woman of character. I rather despised women before. The kind of women I used to meet in my work. They were either over-rich, empty-headed customers or assistants without much will of their own. Agnes has a will, drive, ideas. She'll make a good home for me. Further me in my career. Charm my superiors. Manage my finances. Well, we all have to commit ourselves sometime and I have done just that with Agnes. I know I've taken the right step. I have no doubts about it."

Two days later, there was another letter for Fancy from the redoubtable Agnes. She had been to see the managing director no less of the store where he worked. They were endeavouring to keep Fancy's job open for him, but were under great pressure to fill it, due to the thousands of servicemen becoming available. In the words of the managing director, the labour market was being flooded. Therefore, Agnes concluded, it was imperative that 'her Frederick' obtain early demobilisation. If there was going to be any difficulty, she would come to Calshot and deal with his commanding officer herself.

Hastily Fancy, with the help of the skipper, dealt with the air force authorities and his application for release was granted. This coincided with Taff's departure on a corporal's course, so we spent the evening in a Fawley pub discussing old campaigns like ancient veterans. The Docs came up in our conversation and I decided to try and see them. Next day I made enquiries at station sickquarters. The medical orderly there gave a slow smile. The Docs were indeed around. He scribbled an address and telephone number for me. I should be able to contact them in Southampton, but ring them first the medical orderly advised. I did so and was eventually put through to Doc. Initially he sounded a little vague as to who I was, but he agreed I could visit him the following afternoon. Not for too long. He was frantically busy these days.

The address proved to be the administrative annexe of a large hospital. At the entrance I happened to bump into Mrs Doc, bespectacled and as benign-looking as usual. She wore an outfit vaguely ressembling a nurse's uniform and carried a briefcase. She also had a little car and said she must rush, being due in both Romsey and Winchester that afternoon. I should ask in reception for her husband whose office was on the second floor. Five minutes later I was shaking hands with Doc. He appeared important, perhaps because of his business suit and splendid office with a view of the Solent. Another briefcase stood open on his desk. He too could not give me much time as he was off to Eastleigh and Southsea.

I learned that both Docs were working towards the brave new world of a national health service. They enjoyed wonderful sounding titles and the use of cars, though petrol rationing still applied to most other people. I tried talking about our old boat and crew, about what had happened to Paddy, Fancy, Sparks and Taff, but Doc's eyes kept straying to a wall clock. I took my leave sooner than anticipated and, although I neither saw nor heard of the Docs again, I am sure they did very well for themselves.

Back at Calshot, Timber told me of Skip leaving next morning. His fiancee was coming to collect him by the main gate and our skipper had intimated how he would like to say goodbye to us, the two remaining members of his crew. That Skip had a fiancee was new to Timber and me. We speculated on which of his tall slim sensations it would be. As I approached the main gate next day, I saw Skip, with a couple of suitcases beside him, chatting to Timber. As yet the fiancee had not put in an appearance.

About five minutes later, a limousine came purring down the lane towards the camp and us. It was driven by a chauffeur who opened a rear door for the lady to alight. She was someone Timber and I had not seen previously. Instead of yet another beauty far outshining the rest, this was a dumpy little woman, wearing twin-set and pearls, also a slightly sour expression. Skip kissed her cheek and made the introductions. In a way she was like Doc, not in the least interested. Like him she pointedly checked the time. They would have to hurry if they were to have lunch in Southampton. After they had departed, I remarked to Timber on our skipper having such a plain fiancee after all the lovelies we had seen. Timber's reply was serious rather than cynical.

"Obviously," he said as if instructing me on the facts of life, "they were Skip's pleasures before he got back to business. The business of family commitments. Of property, estates, finances, trusts and an arranged marriage. Two top families getting together I would reckon. Their respective solicitors and accountants sorting out the legal and money sides. Skip and she will have to produce sufficient children to keep the combined line going. Not much love to it, but they'll live comfortable enough lives compared to the rest of us. They'll never have to worry where the next £100,000 is coming from."

Three weeks passed superintending West Indians before I was summoned to the orderly room. Timber was already there and it turned out that he and I were due to be demobilised in London next day. Leaving an RAF camp is, I should imagine, somewhat akin to getting out of prison. One is given a clearance card, then has to go the rounds of the various sections. For example, the sick bay must vouch there is no outstanding treatment and the guardroom there are no unexpurgated crimes. Timber and I kept encountering each other as we criss-crossed the camp. When I saw him towards the end of a weary day, Timber asked me to meet him at the pierhead in an hour's time.

There Timber had already borrowed a marine tender from the duty crew and we set off along the Solent, past the lines of old war-weary boats, swinging idly on their buoys. Ours looked the same as the others until we stepped aboard and Timber unlocked the wheelhouse door with a borrowed key. The bare interior was dispiriting at first, but one soon recalled what it had once been like. I thought of those long hours in the wheelhouse with Timber at the wheel, Skip on the bridge and the rest of us wedged where there was room. As I stood thinking of those air/sea rescue days and nights, Timber took the wheel and gazed at the darkening sea.

"Well," he said at last, "we did save a few lives."

Night was falling fast so we locked up, cast off and returned the borrowed marine tender to the pier. We then had the long walk back from the lower to the upper camp, which we did in silence. At breakfast next morning, colleagues kept pulling my leg about escaping at last. I smiled, but my heart was not in it. Half an hour later, I joined Timber in the van taking us to Southampton Station. From there we could still smell the sea but, once inside the railway compartment, that too was over. Timber and I sat silently opposite each other as the Southampton suburbs were replaced by the Hampshire countryside.

All too soon, the Surrey countryside was taken over by London suburbs. I gazed out at the tentacles of factories, shops and houses from which I had temporarily escaped. From Waterloo Station, Timber let me navigate him to the demobilisation centre in West London. Rumours had reached us at Calshot that the most efficient part of our RAF service would be ending it. Rumour proved correct. Hundreds of airmen went through the system each day to be measured and provided with a complete outfit of civilian clothes. By present standards the cloth was good, though the cut would raise hoots of contemporary laughter. After I had been fitted with my ready-made suit, it was wrapped together with the other items in a neat brown paper parcel. Then we went to clerks in cubicles for our discharge papers and the country's gratitude. Solemnly each of us signed for and received the princely sum of £50. The nation's gratitude worked out at £10 a year.

I had kept my eye on Timber during these operations and we met outside clutching our brown paper parcels. Timber said he hoped to catch the night train to Newcastle so he must rush. Awkwardly, because of the parcels we shook hands. I said,

"Remember Tangier."

"Watch it," he said and was gone.

None of us ever saw each other again.

Appendix A

RAF AIR/SEA RESCUE SERVICE

ORIGIN

The RAF Air/Sea Rescue Service can justifiably place its origin in the same year as the formation of the Royal Air Force — namely 1918. The RAF stemmed from the Royal Flying Corps of World War I fame and the slightly lesser known though no less important Royal Naval Air Service. It was the latter service which contributed a pool of 239 miscellaneous vessels to make up the early marine sections of the RAF.

SEAPLANES

During the 1920's, there was keen aviation interest in seaplanes. These were seen as the aircraft of the future and the thinking ran as follows. Water being flat could be said to provide perfect runways. Therefore rivers and lakes, also the estuaries, inlets and bays of our island coast, offered innumerable ready made airfields, particularly for defence purposes. So seaplanes developed as did the marine craft tending them.

There were of course a great many other types of boat required for a multiplicity of seaplane duties. The seaplanes had to be fuelled and serviced while on the water. Aircrews had to be ferried to and from them. Planes were towed to moorings and to slipways for on-shore working such as repair of damage and engine changes. Above all vessels had to standby while seaplanes took-off and landed. These craft were officially known as safety launches though they inevitably acquired the nickname of crash boats.

Yet the early days of aviation were almost idyllic so far as safety was concerned. The seaplanes took-off and landed at a leizurely 40 mph with the standby boat just about managing 15 mph. Thus the tender was positioned well ahead of the seaplane to be opposite it at the critical moments of lift-off and landing. Accidents were generally not serious and made good stories afterwards in the mess.

SCHNEIDER TROPHY

Technology however cannot remain static. There were many incentives to build faster aircraft and the Schneider Trophy (presented to the Aero Club of France by Jacques Schneider) was one of them. The first contest was held in 1913 and the last in 1931. Their leading particulars are tabulated on the following page because they had far reaching effects on aircraft developments, on defence policies and on the evolution of the RAF Air/Sea Rescue Service.

Year	Aircraft	Winner	Speed (mph)
1913	Deperdussin 160 hp Gnome	M. Prevost France	45.75
1914	Sopwith 100 hp Gnome	C.H. Pixton Britain	86.80
1920	Savoia 550 hp Ansaldo	L. Bologna Italy	107
1921	Macchi 200 hp Fraschini	G.de Briganti Italy	111
1922	Supermarine 450 hp Napier	H.C. Biard Britain	145.70
1923	Curtiss 465 hp Curtiss	Lt.D. Rittenhause United States	177.38
1925	Curtiss 600 hp Curtiss	Lt.J. Doolittle United States	232.68
1926	Macchi 850 hp Fiat	Maj. de Bernardi Italy	248
1927	Supermarine 860 hp Napier	Flt.Lt. Webster Britain	281.68
1929	Supermarine 1800 hp Rolls Royce	F.O. Waghorn Britain	328.63
1931	Supermarine 2300 hp Rolls Royce	Ft.Lt. Boothman Britain	340.08

It will be noted that over an 18 year time span, engine horsepower rose from 160 to 2300 and aircraft speeds from a mere 45 to 340 miles per hour. Thus, within two decades, the fledgling aviation industry was well on its way towards the speed of sound.

The Schneider Trophy contests were to achieve other ends. The table shows that the last three aircraft were not only British but also made by Supermarine. This company went on to create the Spitfire which, with the Rolls Royce Merlin, helped to save Britain before another decade had passed.

Likewise it will be seen that Royal Air Force officers piloted all three winners. By then the RAF dominated the contest which took place at Calshot on the Solent. RAF Calshot was also to be the home base of air/sea rescue high speed launches, powered by sea-going adaptations of those Napier engines used in the Schneider Trophy contests.

AIR/SEA RESCUE DEVELOPMENTS

It will be evident that ever-increasing speeds at take-off and the lengthening of these runs began to highlight shortcomings in safety and rescue requirements. Initially Royal Air Force bases were obliged to borrow fast coastal craft from the Royal Navy but everyone concerned knew this could only be a temporary measure. Clearly the RAF needed swifter and better equipped boats to provide a worthwhile rescue service.

Another factor entered into these calculations. Apart from seaplanes, land based aircraft were beginning to fly far beyond the coastline. Cover had to be provided for them during peacetime exercises and, though few hardly dared to think about it then, in the event of another war. At this stage, during the early 1930's, the legendary figure of T.E. Lawrence — 'Lawrence of Arabia' — enters the air/sea rescue story.

338171 AC1 SHAW

Following his triumphs in the Middle East and disillusionment with post-World War I politics, Lawrence joined the Army then the Royal Air Force as a ranker. At the seaplane base — RAF Mount Batten, Plymouth — he was closely associated with marine craft in general and the development of hard chine hull construction for sea-going, high speed launches in particular.

Lawrence envisaged craft capable of over 30 mph putting to sea in a matter of seconds. At anchor, they would have a draught of under 3 ft. When moving, this draught would be reduced causing the hard chine hull to rise and slide over the surface of the sea. His was a remarkable prediction of the boats destined one day to form the RAF Air/Sea Rescue Service.

HIGH SPEED PROTOTYPES

The trend towards having high speed launches for air/sea rescue work resulted in the British Power Boat Company at Hythe on the Solent producing a prototype in 1932. Designated a 200 Class Seaplane Tender, this boat was claimed to be the fastest of its kind per foot per horsepower in the world. Its specification was as follows:—

200 CLASS SEAPLANE TENDER

Length	37½ ft
No. of engines	2
Make of engines	Power Meadows
Engine power (each)	100 hp
Top speed	30 mph
Construction	Hard chine

Significantly the RAF trials were carried out by Lawrence who moved from Mount Batten to Calshot. His report ran to some eighty pages and his commanding officer thought it a masterpiece of technology. Lawrence dealt with the hull, engines, fuel system, steering, controls, electrical equipment, instruments, maintenance and boat handling under all weather conditions.

These 200 Class Seaplane Tenders proved to be excellent boats and were used by RAF Marine Craft Sections for many years, though mainly within coastal waters. They were considered a first step in the right direction, but not yet the definitive high speed launch air/sea rescue crews were seeking.

NOTE T.E. Lawrence left the RAF at the end of February 1935. On the 15th May of the same year, while riding his Brough motorbike, he swerved to avoid two cyclists and crashed. Lawrence lay unconscious for five days before dying.

Rare photograph of seaplane tender M157 on its way through English country lane to Royal Canadian Air Force. *(Courtesy R.A.F. Museum, Hendon).*

Despite its inherent qualities, the 200 Class Seaplane Tender suffered from several limitations. It had insufficient range and, in particular, was not big enough for prolonged sea searches. Development work continued and in 1936 the first of the 100 Class High Speed Launches, again from Power Boats, arrived at Calshot. Its specification was as follows:—

100 CLASS HIGH SPEED LAUNCH

Length	64 ft
No. of engines	3
Make of engines	Napier Sea Lion
Engine power (each)	500 hp
Top speed	45 mph
Construction	Hard chine
Range	Over 400 miles

The speed and range, also the capacity for accommodating crews and survivors in some comfort, put this new prototype in a class of its own. From then onwards, the 100 Class High Speed Launch's operational capability formed the basis of air/sea rescue planning and activities. There were variants of the main design known as 'Whalebacks' and 'Hants and Dorsets'. The common factor was the type of construction and the layout of the three engines.

RESCUE TECHNIQUES

The Royal Air Force pioneered and practised air/sea rescue techniques, not only in the desire to save human lives, but for other service reasons. The main reason was that aircrew personnel represented lengthy and costly training. Such highly trained people could not be abandoned without every possible effort being made to save their lives if they came down in the sea. Another reason was that of morale. However low the rescue rate was in the beginning, the knowledge that air/sea rescue launches were ready and waiting below proved a significant morale booster.

Rescue techniques improved from elementary precautions to highly sophisticated systems for survival. Early planes began incorporating rudimentary flotation gear. Crews crossing sea areas wore life jackets which were steadily improved. Similarly a succession of improved rubber dinghies for one and two-man fighter aircraft crews, then for whole bomber crews, became available. Rubber dinghies inflated automatically on entering the water and the later types for bomber crews were equippped with sails, rations, radio and weather protection.

THE CHURCHILLIAN EYE

There is a little known World War II story of Winston Churchill being shown round an establishment responsible for the latest developments in air/sea rescue dinghies. A fully inflated dinghy was laid out with all its ancillary equipment for the great man's inspection. Churchill took one look then enquired where were the playing cards. He suggested that crews waiting to be rescued would appreciate cards to help pass away the time. So a pack of cards was included in the kit of every dinghy and rescued crews all reported what a wonderful idea it was to provide them.

SURVIVAL AND RESCUE

During the developments outlined above, the RAF Air/Sea Rescue Service was still small and ill-equipped when war was declared on the 3rd September 1939. Indeed this service was not formed officially until February 1941 though it saved many lives in the meantime. Yet there was a healthy embryo from which the wartime organisation could grow.

In those early days, it was realised but not publicised that aircrew 'down in the drink' had less than a 20% chance of survival. In fact the figure was more of a guess or a starting point for improvement. It was kept hidden from aircrew and the general public, but the implication haunted those concerned with air/sea rescue duties.

Seasoned RAF marine craft personnel, that is those who moved from peacetime preparations to wartime operations, used to recall how their rescue sorties began with high hopes. Long hours, even days later, when not a single fragment of aircraft wreckage had been sighted, everyone knew there was no chance of picking up live survivors.

Nevertheless some searches went on for a week or more. Fresh marine crews would take over from exhausted ones. Aircrews from the same squadron would fly over the search area. The will to save lives was immensely strong and, when failure was finally admitted, spirits sank low. All those engaged in air/sea rescue work became permeated with the same spirit — to do better.

TRAINING PLANS

Plans were made and put into effect on how best to improve the low recovery rate. They fell into the following main categories:—

Training aircrews on the techniques necessary when ditching and abandoning aircraft.

Ensuring the continuance of survival at sea after the abandoned aircraft had sunk.

Improving signal methods by which ditched aircraft and survival dinghies could be located.

Augmenting air/sea rescue services so that aircrew survivors were brought home safely.

As a result of the above, the rescue rate went up to 35% by June 1941 and later that year to 37%. Those responsible however did not think this nearly good enough. During that summer and autumn of 1941, further top level meetings were held to produce a much more comprehensive and efficient air/sea rescue organisation. At that period, two major escalations of the war effort loomed ahead like storm clouds on the horizon. They were:—

> The massive bomber offensive being built up to hit every part of Germany.
>
> The inevitable invasion of Europe from the west to relieve Russia in the east.

Both operations would require a large and well-trained air/sea rescue service. Training therefore became high priority.

TRAINING PROGRAMMES

There were various training programmes for aircrews. These ranged from appointments of air/sea rescue training officers or training teams to Fighter and Bomber Command stations, to the formation of special schools. For

64ft high speed launch with three Napier Sea Lion engines exercising with Lysander aircraft off Portsmouth 1941. This photograph was used in a recruiting poster for ASR Crews — 1940. (Courtesy Group Captain L.R. Flower MBE.MM.)

example, one such establishment originated at Blackpool under the title School of Air/Sea Rescue. Apart from survival at sea, aircrew personnel were taught escape and evasion techniques on land. This school also carried out trials of new equipment.

So far as the marine part of the RAF was concerned, an instructional school which had existed since 1927 was thoroughly reorganised. The school was at Calshot but, no sooner had it been updated, when bombing in the Solent area resulted in a move to Corsewall, near Stranraer in Scotland. At the peak period of the war, as many as 500 marine crew personnel were under training at RAF Corsewall.

RAF CORSEWALL

At RAF Corsewall, there were three courses running simultaneously and designed to provide a supply of trained men for air/sea rescue duties.

Motor Boat Crew	This course covered the care and maintenance of boats, mooring and anchoring, knots and splices, helmsmanship and sea exercises.
2nd Class Coxswain	A more advanced course included complete boat handling, the action of rudders and propellers, towing and refuelling procedures.
1st Class Coxswain	The final course went on to compass swinging, dead reckoning navigation, coastal pilotage and the carrying out of full scale air/sea rescue work.

In general, motor boat crewmen were ordinary seamen (AC1 or AC2), 2nd Class Coxswains were Leading Aircraftsmen or Corporals, and 1st Class Coxswains either Sergeants or Flight Sergeants. At RAF Corsewall there was also a Marine Officers course. There were quite a few ex-Merchant Navy officers and amateur yachtsmen among them, as well as former lifeboatmen with experience in rescue work. The main requirement was to fit together as a crew.

Each air/sea rescue launch crew was headed by a commissioned officer, generally a Flying Officer or Flight Lieutenant, but more popularly known as Skip or the skipper.

The skipper, two coxswains (one 1st and the other 2nd Class) together with two or three deckhands made up the basic boathandling side of an air/sea rescue crew. They steered and navigated the boat, also manned the guns. No less important were crew members specially trained as medical orderlies, wireless operators and marine engineers. Their principal duties were as follows though at sea and especially on rescue work everyone did everything.

Medical Orderly A medical orderly was attached to each rescue launch engaged on actual operations. He was responsible for the sick bay, medical kit and treatment of survivors.

Wireless Operator This operator played a key communications role, receiving and transmitting messages. He also operated direction-location equipment and was the ship's electrician.

Marine Engineers About one-third the length of a high speed launch was taken up by three or more main engines, auxilliary engines, fuel tanks and other equipment, so there were two engineers in each crew. Most were trained at RAF Locking.

RAF LOCKING

Marine engineers for the RAF Air/Sea Rescue Service, also some of the wireless operators, were trained at RAF Locking, near Weston-super-Mare. This was a permanent air force station and the standard of instruction given there superlative despite wartime conditions. Every course was in the best technical traditions of RAF Training Command — that is, pared to essentials, then put over positively with a careful balance between theory and practice.

The Marine Engineering course at Locking lasted nine months despite urgent needs at coastal bases. It commenced with the theory of engines followed by 2-3 weeks on every type of power plant fitted to RAF marine craft. These covered small outboard motors for runabout dinghies, diesel engines for tenders, up to the 500 hp Napier Sea Lion petrol engines on high speed launches. Later came courses on 1200 hp engines for long range rescue craft.

Tests were either weekly or fortnightly and every trainee had to pass each stage in order to reach the next. Those who failed were returned to their units or relegated to general duties. At the end of the course came a complete examination which determined future rank and pay. At last, marine engineers trained to high standards were posted to air/sea rescue boats and bases.

With the various training programmes outlined above, the RAF was able to assemble crews for the rescue tasks that lay ahead. Over the same period however, a parallel problem existed in the lack of sufficient boats.

COVERAGE AND RESOURCES

From early in World War II, coastal waters surrounding the British Isles were divided into air/sea rescue zones. Each zone had headquarters in contact with all available marine craft. Whenever aircraft from Fighter, Bomber and Coastal Commands were to fly over the sea, they informed air/sea rescue whereupon launches took up positions under the air routes. This organisation was to continue for the remainder of hostilities and well into the post-war years until replaced by present day helicopters.

When the 100 Class High Speed Launch, as previously mentioned, reached the RAF in 1936, 15 more were ordered. Two years later there were still only 9 rescue launches in service and a further two years elapsed before this insufficient number rose to 18. By then the Germans had launched their highly successful Spring offensive of 1940 and conquered France. By then also RAF seaplane tenders from Calshot and other southern bases slipped away to Dunkirk where they were able to rescue some 500 British soldiers.

Further high speed launches were on order and makers of marine craft worked round the clock to build up an adequate rescue fleet. Yet the supply never seemed to catch up with demands. Such demands were initially created by fighter sweeps over enemy occupied France followed by bomber raids on

(Courtesy R.A.F. Museum, Hendon).

Whitley P1502 immediately before ditching. 47° 35'N 08° 38'W

Whitley P1502 ditched, crew taking to dinghy to await rescue (Courtesy R.A.F. Museum, Hendon).

the German heartland. By early 1944, when the long planned and awaited invasion of Europe was nearing reality, the effective air/sea rescue strength was:—

High speed launches	130	182
Seaplane tenders	27	
General purpose pinnaces	25	

RESCUE BOAT ARMAMENTS

Air/Sea rescue crews did not like the idea of mounting guns on their boats, being primarily concerned with saving instead of taking lives. In the early days of World War II, a certain amount of chivalry existed between British and German rescue crews, as did among the fighter pilots of both sides. In the air the object was to win the dogfight, then break off hostilities as the loser parachuted down to earth or into the sea. At that stage, some rescue launches had one or two light machine guns mounted on a convenient part of the deck — just in case.

The case came in August 1941 when the Germans attacked two RAF high speed launches, sinking them both. This sent a shock wave through the rescue service. It resulted in the light machine guns being replaced by twin Brownings mounted in two turrets fore and aft. These turrets were hand not power operated and incorporated an interesting feature. Curved rails ran round them to prevent guns being fired in certain directions. Putting it another way, the rails stopped gunners from blasting parts of their own ship and even fellow crew members.

Following the ill-fated Dieppe raid in August 1942 when, apart from the main casualties, three RAF air/sea rescue launches were destroyed with the loss of 20 crewmen, it was decided to fit yet heavier guns on high speed boats. The extra guns were 20 mm Oerlikon cannon supplied by the Royal Navy for use on RAF launches. Regular gun drills were introduced, with the notice for them bearing the inscription:—

"Here lies the body of Sergeant Keen,
A 1st Class Coxswain of high esteem.
An excellent seaman he was in his prime,
But he thought gunnery a waste of time."

In addition to the regrettable fitting of machine guns and cannon, RAF Air/Sea Rescue boats began to acquire flak mats and armour plating. These were put round wheelhouses so as to protect steering and wireless equipment, as well as crew members grouped there while on operations. Likewise the

High speed launch 2550 on rescue work. Note many guns fitted, also engine room ventilators down aft. *(Courtesy R.A.F. Museum, Hendon).*

engineer down aft was provided with a thick sheet of steel for the back of his seat. Although not common, there were incidents involving damage, casualties and loss of lives to crews trying to save lives. On two occasions, air/sea rescue crews were captured. Thirteen boats in total were sunk due to enemy action.

IMPROVEMENTS IN OPERATIONS

As previously mentioned, early air/sea rescue operations were haphazard affairs involving long hours at sea with no clear idea of what had initially happened or where the survivors (if any) were to be found. Sometimes launches searching for the crew of one aircraft would come upon those of another. A boat would unexpectedly find survivors and bring them safely to harbour leaving another boat still searching the now empty area. Those were the days when survivor recovery rates were far too low.

Such conditions of 'organised chaos' as airmen called them were steadily improved due to crew keenness and especially the ever high standard of RAF training. From 1942 onwards, the Air/Sea Rescue service began to take part in much more precise operations. Significantly these started in the Dover Strait zone where the proximity of enemy-occupied France demanded close knit air and sea liaison. Many pilots who baled out over the Channel were rescued within an hour, if not minutes, of leaving their aircraft.

By this period, the massive offensive by Bomber Command was being mounted and would soon be increased by that of the US Army Air Force. Bombers suffered extensive damage and heavy casualty rates due to ground fire and fighter attacks. The ditching rate off South East England in particular demanded the presence of every air/sea rescue boat that could be made available. It also meant crews spending long hours on a continuous basis in the inhospitable North Sea.

OPERATION OVERLORD

Planning and training for the invasion of Europe had to be thorough as everyone knew it must not fail. The amount of organisation required was immense and, although the RAF Air/Sea Rescue Service had a relatively minor role, it was nevertheless a necessary one. New boats became available for freshly trained crews while most of the existing craft were completely refitted during the early months of 1944. In addition to new boats and the refitting of old ones, extra bases were set up along the South Coast which virtually became an armed camp. By May, all was ready and air/sea rescue crews waited like everyone else.

On D-Day itself, 136 RAF rescue launches were deployed in and about the assault area. During that 'longest day', many pick-ups were made. By the end of D-Day plus nine, these high speed launches had saved the lives of 163 airmen together with some 60 other casualties at sea. Air/sea coverage continued day and night throughout that fateful June of 1944. By the end of the invasion month, no less than 355 lives had been saved.

ARNHEM AND NIGMEGEN

Equally effective were the efforts of air/sea rescue craft and crews deployed later that year to cover airborne attacks on the bridges at Arnhem and Nigmegen. Mobile units operated from Calshot and established bases on the French coast as early as possible. The RAF flew the combat troops there and RAF Air/Sea Rescue waited below them. During those four days in September 1944, 181 survivors were recovered from the sea — 91 in one day.

OTHER THEATRES OF WAR

While the seas round Britain absorbed the main efforts of the RAF rescue service, there were numerous recovery operations taking place in other theatres of war. Pride of place should be given to the lone but gallant HSL 107 that operated out of Malta in the early years of the war together with a couple of seaplane tenders. Later many more marine craft and aircraft were deployed in search and rescue roles throughout the Mediterranean.

In the Aegean, RAF launches were used to cover many operations including landings on the island of Rhodes. These craft also operated round the Greek Archipelago and further east off the Turkish Coast in little known yet desperate actions when nothing seemed to be going right for the British. Then as the tide of war turned, air/sea rescue support was automatically included in each successive Anglo-American operation.

For example, during the Sicilian campaign preceding invasion of the Italian mainland, 45 lives were saved by the RAF Air/Sea Rescue Service taking part. Coincidental with this operation, many aircraft and several rescue craft searched for four days and nights on another emergency exercise. They recovered 42 survivors from Flying Fortresses bound for North Africa following a raid on Regensburg in Southern Germany. Air/sea rescue suppport continued during the Italian campaign and later in Southern France.

Two other areas were covered by air/sea rescue. Such support was never sufficient due to over-extended resources, but lives continued to be saved. Seaplanes and marine craft operating from West African bases saved 200 out of 270 airmen reported down in the Atlantic. In the Far East, from June 1943 until the end of the war, air/sea rescue saved 327 airmen out of some 700 recorded.

LONG RANGE AIR/SEA RESCUE

As the war in Europe drew near to an end, the RAF turned its full attention to the Far Eastern theatre, including air/sea rescue in their plans. Distances between islands, also heavier seas in the Indian and Pacific Oceans, demanded larger, more powerful launches with a greater range than those of existing craft. The quickest way to achieve this was to convert ex-Navy boats, also to re-train existing crews.

The conversion of boats was achieved as follows. Forty Fairmile D-Type craft, formerly used as motor torpedo boats by the Royal Navy, were refitted for Royal Air Force air/sea rescue usage. These boats were 115 ft long, powered by four Packard 1200 hp engines (derived from Liberty engines) and with supplementary fuel tanks had an operating range up to 1500 miles. Conversions were quickly carried out at every dockyard having spare capacity and the air/sea rescue crews came aboard, after their re-training courses, to help with the commissioning work.

These long range boats were painted in the traditional RAF air/sea rescue colours and numbered 001, 002 upwards. Squadrons of eight boats at a time were then assembled at RAF Calshot and then on to Mount Batten. The initial leg of their long journey to the Far East was across the Bay of Biscay and round the Iberian Peninsula to Gibraltar. The first squadron had reached Egypt and the second entered the Mediterranean when atoms bombs fell on Hiroshima and Nagasaki in August 1945. The swift surrender of the Japanese meant that World War II was over and the requirement for air/sea rescue much diminished.

POST WAR ACTIVITIES

The peacetime RAF, also NATO, made increasing use of sea ranges for practice bombing, gunnery and rocket-firing. This put the aircrews taking part at some risk. There was also the switchover from sea to air trooping which likewise required some kind of rescue service. New conflicts such as the Korean War arose and that again demanded rescue support. However the type of support began to change.

During this period, the use of aircraft instead of marine craft for sea searches increased with helicopters taking over the leading role. By 1957, helicopters were predominant and since then, while air/sea rescue work has continued uninterrupted, the pattern of operations is quite different. Location and rescue is much swifter. Within minutes of an accident, the versatile helicopter is on the scene, operating above the roughest seas. Nevertheless the lessons learned over the years are embodied in the survival techniques practised by modern aircrews.

CONCLUSION

The following figures are the official figures. By the end of World War II, some 300 boats and crews of the RAF Air/Sea Rescue Service had been responsible for saving 14,000 lives. Out of that impressive total, 8000 were aircrew. Apart from them not meeting their deaths in the inhospitable sea, their long and costly training was not wasted. It has been said that, had only a tenth of the total been saved, this would still have more than justified the air/sea rescue programme.

Little is known about the activities of the RAF Air/Sea Rescue Service, hence this book and especially these appendices. The boats were manned by very ordinary people who were far from being seamen in the true sense of the word. The Royal Navy creamed off the professionals and the Royal Air Force did its best with the amateurs. Yet their sustained efforts and the diversity of their operations demonstrated the British flair for improvisation.

Finally, if there was one aspect which exemplified the spirit of those RAF air/sea rescue crews, it was the universal desire to save human lives. Crews on search would continue beyond exhaustion to the last drop of fuel if there was the slightest chance of finding a single survivor. They practised rather than preached the motto of their service — 'The sea shall not have them'.

RAF HIGH SPEED LAUNCHES

GENERAL

A newcomer to an RAF Air/Sea Rescue Base in wartime, especially to such big bases as Mount Batten or Calshot, could be excused for being bewildered by the multiplicity of boats in all shapes and sizes. All would be painted black with grey decks and superstructures, the bows bearing RAF roundels on either side and the RAF ensign flying at the masthead. They made a fascinating if confusing sight.

Further observation would eliminate the smaller boats such as planing dinghies and marine tenders, also service vessels like refuellers. Sooner or later one's eye is drawn to the low rakish launches on which there are letdown crash nets, life rafts and direction finding gear as well as guns, cannon and flak mats. These the observer would guess correctly must be the High Speed Launches of the RAF Air/Sea Rescue Service.

WHALEBACKS

Known as Whalebacks because of their distinctively curved hulls and humped cabins, these boats were introduced in the RAF Air/Sea Rescue Service in 1940. They saw action in the Battle of Britain and, as their production got underway, were the main rescue launches operating in the English Channel and the North Sea. Similar craft served in the Mediterranean and the Indian Ocean. Some Whalebacks were in continuous operation throughout World War II.

The Whalebacks were 63 ft in length with a beam of 17ft 2″. Powered by three Napier Sea Lion 500 hp petrol engines, they were capable of speeds over 35 knots and had a range of about 420 miles. After several boats were attacked, their initial armament of one or two light machine guns was increased to a pair of twin gun turrets and a 20 mm cannon.

The Whaleback design stemmed directly from the family of fast hard chine boats conceived by Hubert Scott-Paine. These were detail designed by George Selman and built by the British Power Boat Company at Hythe on the Solent. Sixty-nine were produced for the RAF between 1940 and 1942. Similar looking but larger boats were constructed for the Royal Navy and the US Navy — the latter being made by the Electric Boat Company under licence from the UK.

These launches pioneered the use of resin bonded wooden structures and the Whaleback incorporated the very best of these techniques. Resin bonding

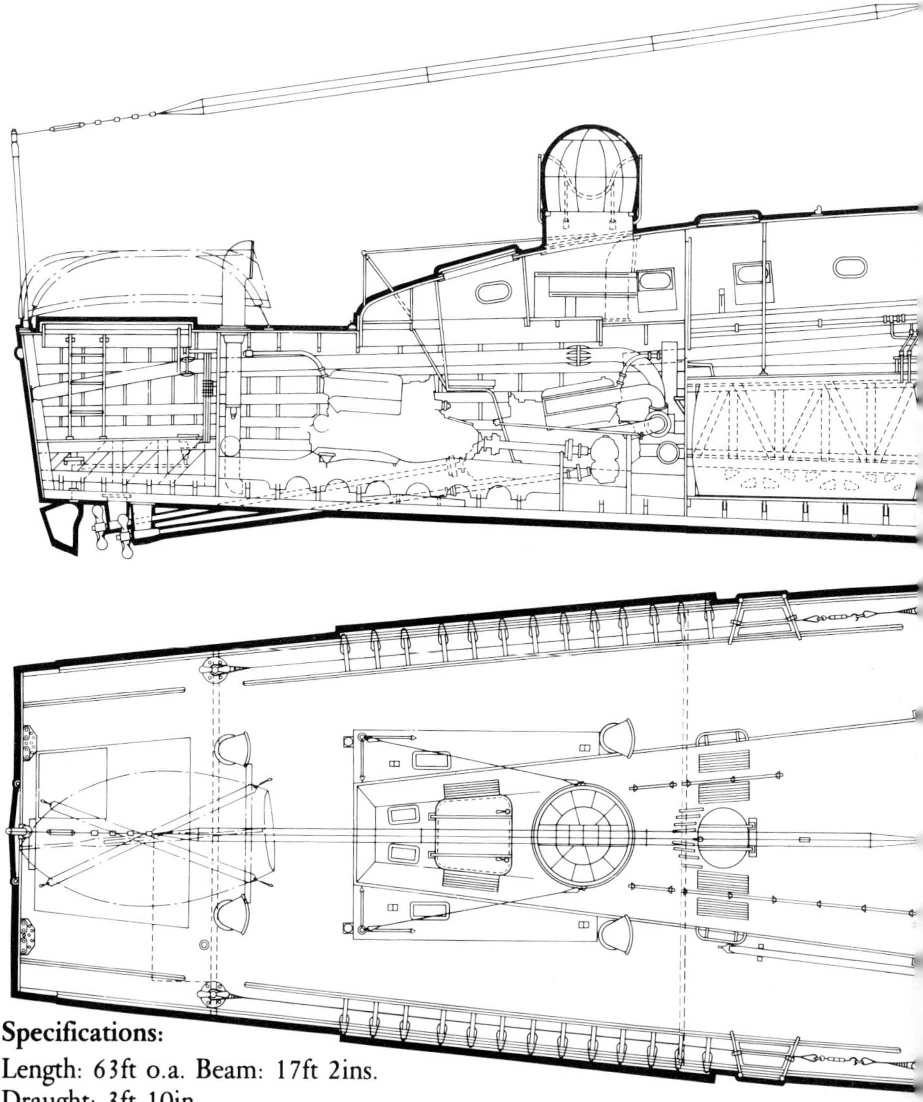

Specifications:

Length: 63ft o.a. Beam: 17ft 2ins.
Draught: 3ft 10in.
Power: 3 Napier Sea Lion petrol engines of 500hp each.
Speed: 35 knots (max. for ¼hour). 28 knots (maximum continuous).
Range: 420 nautical miles.

63ft WHALEBACK
(Boat 14

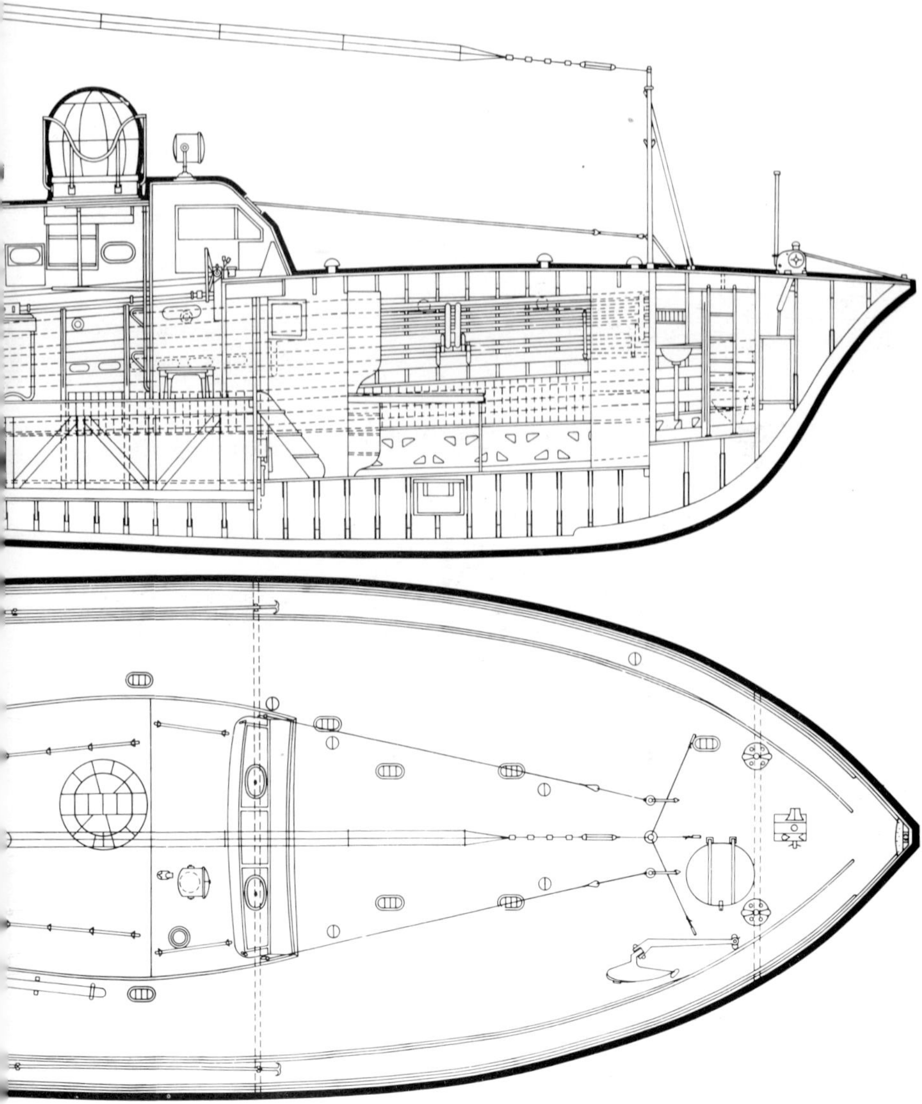

(Less armaments)

I SPEED LAUNCH
onwards)

was used in lateral and longitudinal frames, stems and keels, also for wheelhouses and cabins. The frames were closely spaced for added strength and these features, combined with the double diagonal hull skins, produced craft capable of riding the roughest seas.

Whaleback rides are best recalled than experienced. It was claimed that these boats could take far more than their crews. Notwithstanding, air/sea rescue crews endeavoured to be as tough as their boats. They insisted on going out in all weathers, eventually gaining grudging admiration from the Royal Navy. Whalebacks were to the rescue service what Spitfires were to Fighter Command. The table below lists where they operated.

RAF BOAT No.	ENTERED SERVICE	BASED AT OR BASE No.	STRUCK OFF CHARGE
122	21.1.41	Calshot, Dover. Lost by enemy action.	24.8.42
123	5.2.41	Calshot, Dover, Calshot, Dover. Lost by enemy action.	24.8.42
124	13.3.41	Calshot, Great Yarmouth, Acklington-Blyth, Lucas-Tayport, Littlehampton, Grimsby, Felixstowe, Gorleston.	15.5.46
125	21.4.41	Dover-Ferrydock, Coltishall, Felixstowe, Brighton-Newhaven, Gorleston, Loch Boysdale.	
126	5.5.41	Calshot, Newhaven, Torry-Aberdeen, Littlehampton, Felixstowe, Weymouth.	18.6.46
127	6.41	Calshot, Dover, Ramsgate. Damaged in collision.	7.5.46
128	22.8.41	Calshot, Malta — Transferred to the Italian Air Force.	17.9.47
129	2.7.47	Calshot, Malta. Lost by enemy action.	26.4.42
130	30.8.42	Calshot, Gorleston, Tayport, Littlehampton, Grimbsy.	16.7.46
131	19.9.41	Calshot, Dover, Ramsgate, Grimsby, Acklington-Blyth, Tayport, Littlehampton, Felixstowe. Lost enemy action.	1.45
132	16.9.41	Calshot, Acklington-Blyth, Yarmouth, Gorleston, Dover-Ferrydock.	11.4.45
133	30.4.41	Calshot, Blyth, Dyce, Chittagong, Bone, Algiers-Bougie, Algiers-254. Transferred to the Italian Air Force.	29.4.47
134	.41	Calshot, Felixstowe, Newhaven, Dover-Ferrydock.	16.7.46
135	21.10.41	Calshot, Ballei-Holyhead, Greenock, Tenby. Lost by enemy action.	15.10.42
136	5.11.41	Calshot, Holyhead, India, Colombo, Calcutta, Chittagong.	19.11.46

137	12.11.41	Calshot, Torry, Shappinsay-Orkneys, Calcutta, Bone, Algiers-Bougie, Italy.	5.47
138	28.11.41	Calshot, Dover. Lost by enemy action.	28.7.42
139	1.1.42	Calshot, Grimsby, Middle East-Philippville, Algiers, Port Said.	17.9.47
140	42	Calshot, Dover, Ramsgate, Corsewall.	14.5.46
141	18.10.40	Calshot, Grimsby, Port Said.	2.48
142	15.11.42	Calshot, Grimsby, Dover, Calshot, Gibraltar, Felixstowe.	25.10.46
143	24.11.40	Calshot, Ramsgate, Dover. Lost by enemy action.	5.41
144	24.3.41	Calshot, Grimsby, Tenby, Gorleston, Ford.	21.11.46
145	5.3.46	Calshot, Newhaven, Felixstowe.	5.3.46
146	26.5.41	Calshot, Acklington-Blyth, Chittagong, Bone. Badly damaged by enemy action........written off.	6.12.46
147	31.5.41	Calshot, Dover. Lost by enemy action.	24.8.42
148	24.8.41	Calshot. Lost in transit on way to Malta on the Imperial Star by enemy action.	10.10.41
149	13.8.41	Calshot, Ramsgate, Dover, Felixstowe, Thorney Island, Fowey.	7.5.57
156	22.1.42	Calshot, Newhaven, Tenby, Fleetwood, 213.	10.10.46
157	1.1.42	Acklington-Blyth, Torry-Aberdeen, Blyth, Dover-Ferrydock, Grimsby.	24.12.46

Rescue launch 122 at half speed. *(W.D. Pereira collection)*

RAF BOAT No.	ENTERED SERVICE	BASED AT OR BASE No.	STRUCK OFF CHARGE
158	9.1.42	Calshot, Blyth, Yarmouth, Gorleston, Dover-Ferrydock, 1101, Fowey.	11.2.47
159	18.2.42	Calshot, Alexandria (207-Jan '44).	4.47
160	23.3.42	Calshot, Alexandria.	———
161	23.2.42	Calshot, Alexandria. Sunk by enemy action.	30.11.44
162	23.2.42	Calshot, Alexandria.	———
163	19.1.42	Calshot, Columbo, Koggala.	7.46
164	22.1.42	Calshot, Far East/India. Reduced to produce.	9.45
165	22.1.42	Calshot, Alexandria.	2.48
166	19.3.42	Calshot, Alexandria, Malta, Alexandria.	8.48
167	18.2.42	Calshot, Tenby, Pembroke. Lost by enemy action.	19.9.42
168	23.3.42	Calshot, Lerwick, Felixstowe, Newhaven, Gorleston, Weymouth, Warmwell, Converted to remote control target launch.	31.1.50
169	24.3.42	Calshot, Newhaven, Ramsgate, Dover-Ferrydock, 62MU.	16.7.46
170	24.3.42.	Calshot, India/Columbo.	27.10
171	24.3.42	Calshot, Newhaven, Gorleston, Mid. East-Philippville, Algiers-Bougie, Algiers-254, Italy, Transferred to the Italian Air Force.	29.4.47

Rescue Craft 128 underway. The photograph clearly shows why launches of this type were nicknamed 'Whalebacks'. *(Courtesy R.A.F. Museum, Hendon).*

63' HSL. 130 in the English Channel/Dover Strait area.

(Courtesy Group Captain L.R. Flower MBE.MM.)

RAF BOAT No.	ENTERED SERVICE	BASED AT OR BASE No.	STRUCK OFF CHARGE
172	26.4.42	Calshot, Buckie, Chittagong, Bone, Italy. Converted to Target Towing Launch.	8.48
173	26.4.42	Calshot, Lerwick, Shappinsay, Felixstowe, Newhaven, Fleetwood.	11.4.45
174	16.5.42	Calshot, Stornoway, Bone, Italy, Algiers, Italy. Conversion to Target Towing Launch.	26.4.46
175	16.5.42	Calshot, Stornoway, Bone, Algiers-251, Algiers-254. Transferred to the Italian Air Force.	29.4.47
176	16.5.42	Calshot, Greenock, Campbeltown, Chittagong, Bone, Algiers-251 and Bone.	7.47
177	3.6.42	Calshot, Newhaven. To Chichester Yacht Co.	7.5.46
178	3.6.42	Calshot, Newhaven, Ramsgate, Dover, 85MU, Felixstowe.	19.12.44
179	3.6.42	Calshot, Fleetwood, Middle East-Philippville, Bougie, Algiers-251 & Bougie, Italy, Algiers. Transferred to the Italian Air Force.	29.4.47
180	3.6.42	Calshot, Tenby, Dover-Ferrydock, Ramsgate, Gorleston, Calshot.	5.5.46
181	6.7.42	Calshot, Falmouth, Gibraltar, 62MU.	15.5.46

Dramatic shot of rescue craft 142 at top speed and well out of the water as it comes off the top of a wave. Note: Partly stripped-down gun turrets. *(Courtesy R.A.F. Museum, Hendon).*

RAF BOAT No.	ENTERED SERVICE	BASED AT OR BASE No.	STRUCK OFF CHARGE
182	3.7.42	Calshot, Campbeltown, Chittagong, Bone, Malta.	5.5.44
183	31.3.43	Calshot, Gosport, Newhaven, Ramsgate.	18.6.46
184	10.12.43	Calshot, Gosport, Gorleston, Weymouth, Warmwell, 84MU, 62MU.	18.6.46
185	8.10.43	Calshot, Great Yarmouth, Blyth, Gorleston, Weymouth. To Chichester Yacht Co.	7.5.46
186	6.7.42	Calshot, Gosport, Newhaven, Calshot	5.3.46
187	6.7.42	Calshot, Yarmouth, 62MU, Calcutta, Algiers-Bougie.	15.11.45
188	6.7.42	Calshot, Torquay, Grimsby-Coates, 62MU, Calcutta, Algiers 254 and 251, Malta.	31.8.47
189	15.8.42	Mount Batten, Dover-Ferrydock, Ramsgate, 64MU, 88MU.	16.7.46
190	25.8.42	Calshot, Newhaven, Calshot, 62MU.	15.5.46
2546	4.9.42	Calshot, Fleetwood, Middle East-Philippville, Algiers-Bougie, Algiers-254. Transferred to the Italian Air Force.	29.4.47
2547	19.8.42	Calshot, Fleetwood, Middle East-Philippville, Algiers-Bougie and 254.	16.12.46.
2548	28.8.42	Calshot, Newhaven, Dover-Ferrydock, Ramsgate, 62MU, Drummore-Ayr, 213MU, MC Construction Dumbarton.	11.9.46
2549	11.9.42	Calshot, Leuchars, Dover, Ramsgate, Dover-Ferrydock, 34ASR, Ramsgate, 62MU, Weymouth, Calshot, 62MU.	15.5.46
2550	13.11.42	Calshot, Leuchars, Gorleston, Felixstowe, Tenby.	18.6.46
2551	13.11.42	Calshot, Leuchars, Gorleston. Lost by enemy action.	29.6.44

Two rescue boats on patrol. View over rear turret shows the typical wake of high speed launches. (Courtesy R.A.F. Museum, Hendon).

Wartime air/sea rescue launch 177 on operations with twin-Browning gun turret and 20mm Oerlikon cannon at stern manned. (Courtesy R.A.F. Museum, Hendon).

Rare aerial view of early wartime rescue launch in action. Note numbers 181 on foredeck as well as on bow.

Rescue launch 185 at high speed with skipper on bridge and a man at the stern cannon. *(Courtesy R.A.F. Museum, Hendon).*

HANTS AND DORSETS

The Hants and Dorsets entered RAF service in December 1942 when the first production boat was collected from Hythe and taken to Calshot for trials. In March 1943 it moved to Newlyn where, with its sister ship, a new air/sea rescue base was formed. Eighty-nine Hants and Dorsets were built, mostly during 1943. British Power Boats was then making other boats for the RAF and the Royal Navy at a production rate of 12 a month.

Hants and Dorset launches provided sufficient space below deck for crews to live aboard and to operate for long periods away from their bases. This type of craft therefore helped to increase rescue coverage around the coasts of Britain and in the Mediterranean. They were used extensively during the

Launch 2626 as featured on front cover showing its paces. *(Courtesy R.A.F. Museum, Hendon).*

invasion of Europe and after D-Day operated out of liberated Channel ports. Hants and Dorsets likewise equipped several air/sea bases in the Indian Ocean.

RAF BOAT No.	ENTERED SERVICE	BASED AT OR BASE No.	STRUCK OFF CHARGE
2552	23.11.42	Calshot, Newlyn, Predannock-929 Sqdn., Falmouth, Newlyn, Calshot, AEAF 85GP, Aberdeen-Torry, Newhaven, 238MU, Gibraltar, 238MU, 103MCU, 1110MCU, 238MU, 1107MCU, 1113MCU, D of NC.	4.58
2553	8.12.43	Calshot, Newlyn, Predannock, Falmouth, Newlyn, Calshot, 85ASR, Aberdeen-Torry, Newhaven, 85MU, 238MU. Converted to Target Towing Launch, 1107MCU, Newhaven, 238MU, FEAF.	11.12.51
2554	11.1.43	Calshot, Newlyn, Predannock, Falmouth, Newlyn, Calshot, AEAF 85GP, Aberdeen-Torry, Calshot, 238MU, C.Cmd. 1103NCU Pembroke Dock, Calshot. Converted to Target Towing Launch, 238MU Gibraltar.	18.3.58
2555	11.3.43	Calshot, Gosport, Felixstowe, Gt. Yarmouth, Gorleston, Weymouth, 84MU, 238MU. Converted to Target Towing Launch, sunk-Invergordon	15.1.52
2556	11.1.43	Calshot, Newlyn, Predannock, Falmouth, Newlyn, Calshot, AEAF, 85GP, Gorleston, Calshot, Dover-Ferrydock, Ramsgate, 238MU, 62MU. Converted to Target Towing Launch, 1107MCU, 238MU, MCTS Instructional.	24.4.54
2557	19.1.43	Calshot, Felixstowe, 62MU, 238MU. Converted to Target Towing Launch, Newhaven, 1107MCU, 238MU, 1100MCU Alness.	4.58
2558	27.1.43	Calshot, Felixstowe, 62MU, 51EU Bombay, Columbo. Converted to Target Towing Launch.	.49?
2559	15.2.43	Calshot, North Coates, Calshot, AEAF 85GP, Calshot; Isle of Man, Andreas. Converted to Target Towing Launch, Felixstowe, 1107MCU.	19.10.56
2560	17.2.43	Calshot, North Coates, Calshot, AEAF 85GP, Gorleston, Lowestoft, Calshot, Dover-Ferrydock, Ramsgate, 238MU, 62MU. Converted to Target Towing Launch, 238MU, RAF Selatar.	10.57

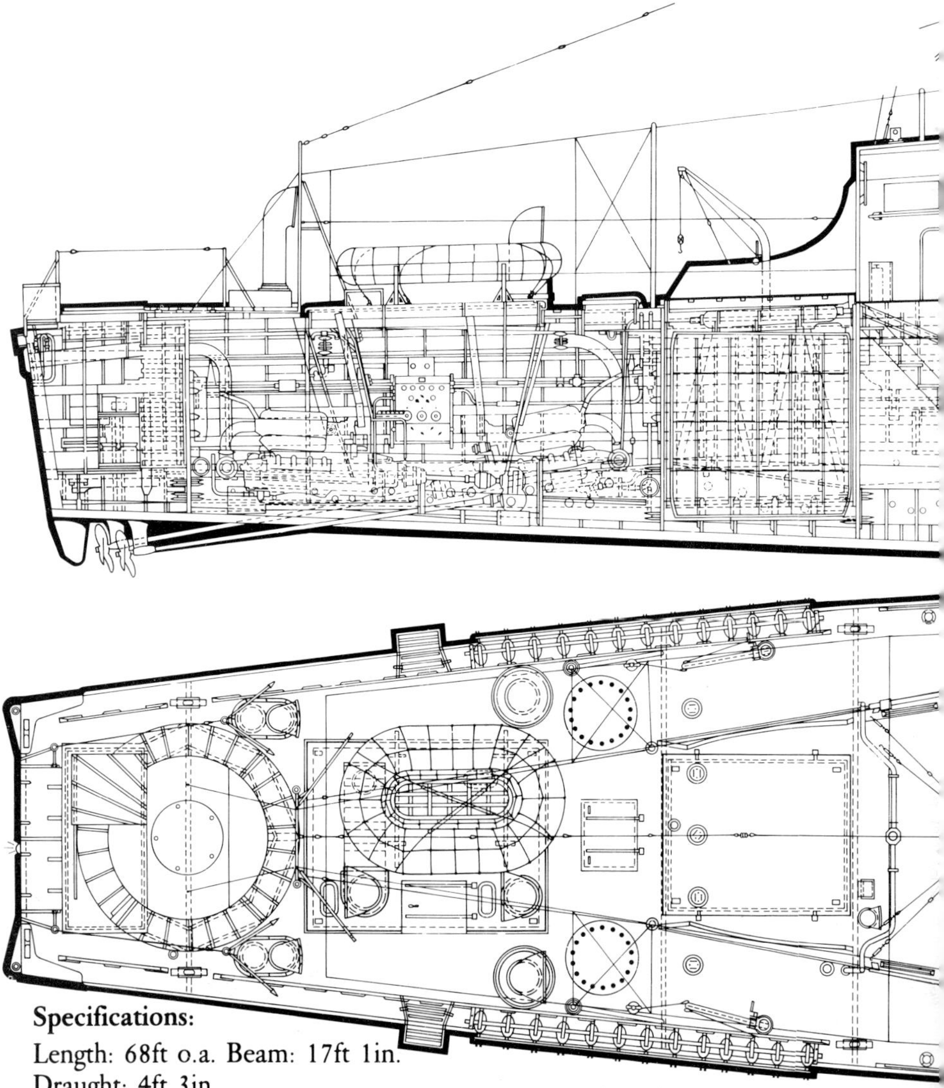

Specifications:

Length: 68ft o.a. Beam: 17ft 1in.
Draught: 4ft 3in.
Power: 3 Napier Sea Lion petrol engines of 500hp each.
Speed: 28.5 knots (max ½hour). 23.5 knots (max. continuous).
Range: 580 nautical miles.

68ft HANTS & DORSET HIGH SPEED LAUNCH

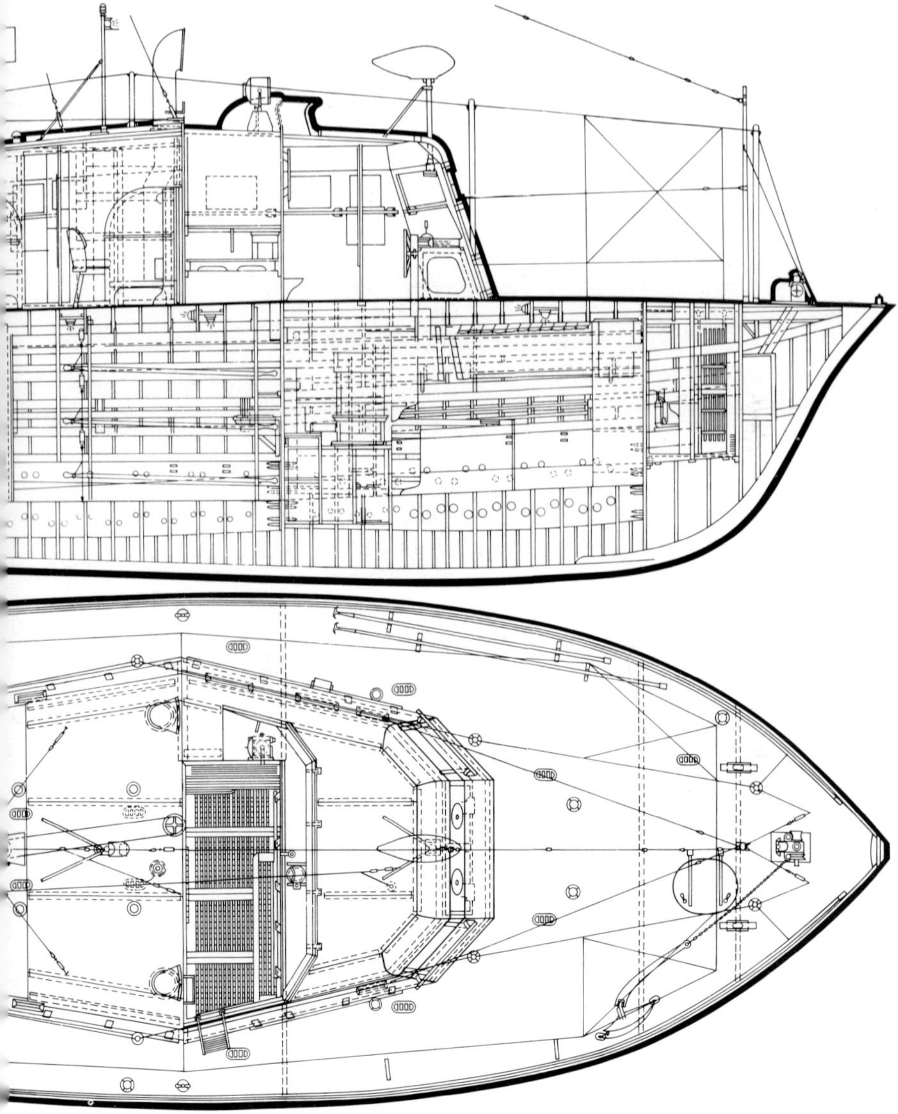

Boat 2320 and onwards)

RAF BOAT No.	ENTERED SERVICE	BASED AT OR BASE No.	STRUCK OFF CHARGE
2561	17.2.43	Calshot, Gosport, Falmouth, Calshot, AEAF 85GP, Calshot, Tayport, 238MU. Converted to Target Towing Launch, 1103MCU, Felixstowe, 238MU, D of NC.	29.10.58
2562	24.2.43	Calshot, Felixstowe, AEAF 85GP, Calshot, 62MU, 129 Staging Post, Cocos Islands.	
2579	26.3.43	Calshot, North Coates, Calshot, AEAF 85GP, Calshot, Weymouth, Pembroke Dock. Converted to Target Towing Launch, 1101 MCU, Felixstowe, D of NC.	17.7.59
2580	5.3.43	Calshot, 62MU, Algiers-254ASR, Bone, Algiers.	16.3.48
2581	18.3.43	62MU, Calshot, Algiers-254ASR, 253ASR, Bone, 253ASR.	11.3.48
2582	18.3.43	62MU, Calshot, Algiers-254, Algiers, 213MU. Damaged beyond repair.	3.3.47
2593	31.3.43	62MU, Calshot, Algiers-254, Italy, Algiers, Italy. Transferred to Italian Air Force.	29.4.47
2594	31.3.4	Calshot, North Coates, Calshot, AEAF 85GP, Gorleston, Lowestoft, 43GP, 85MU, Calshot, Weymouth, Tenby, Pembroke Dock. Converted to Target Towing Launch, Calshot, MCTS Mountbatten.	9.57
2595	21.4.43	Calshot, Algiers.	11.3.48
2596	21.4.43	Calshot, Algiers-254, Malta, Italy, Algiers-251, 238 MU.	16.3.48
2597	21.4.43	Calshot, Algiers, Italy, Algiers-254.	11.3.48
2598	21.4.43	Calshot, Italy, Algiers. Lost by enemy action.	13.11.43
2599	21.43	Calshot, Italy, Algiers, Alexandria, 238MU, 1608MCU.	11.3.48
2600	19.5.43	Calshot, Algiers-254, Italy, Transferred to the Italian Air Force.	1.47
2601	19.5.43	Calshot, Algiers-254, Algiers-Bone, Algiers-Bougie, Italy, 238MU.	16.3.48
2602	19.5.43	Calshot, 238MU, 212ASR, Alexandria, MEAF. Converted to Target Towing Launch.	———
2603	7.6.43	238MU, 212ASR, Alexandria, MEAF, Converted to Target Towing Launch. Burned.	15.9.54
2604	4.6.43	238MU, 212ASR, Alexandria. Converted to Target Towing Launch.	———
2605	4.6.43	238MU, Malta, Algiers-254, Italy, Algiers-254, 238MU.	16.3.48
2606	4.6.43	238MU, Malta, Bone, Malta, Algiers-Bougie, Italy, 238MU.	16.3.48

RAF BOAT No.	ENTERED SERVICE	BASED AT OR BASE No.	STRUCK OFF CHARGE
2619	4.6.43	Calshot, 238MU, Madras, 231ASR, Ferryside.	7.5.48
2620	21.6.43	238MU, Calshot, 205 Malta, Bone, Malta, Algiers-254, Bougie, Italy, 238MU.	16.3.48
2621	21.6.43	238MU, 205 Malta, Bone, Algiers-254.	11.3.48
2622	21.6.43	238MU, 213ASR Middle East, Alexandria- 207.	11.3.48
2623	21.6.43	238MU, Mombassa, China Bay, 237 Columbo, 54EU, Calcutta, 229 Calcutta.	23.9.46
2624	20.2.47	238MU, 210ASR, 237 Columbo, Trincomalee, 235 China Bay.	20.2.47
2625	1.7.43	238MU, 219ASR, 210ASR, Alexandria. Converted to Target Towing Launch.	27.11.59
	1.7.43	238MU, 219ASR, 210ASR, Alexandria. Converted to Target Towing Launch.	27.11.59
2626	1.7.43	238MU, Vizagapatam, Bombay, ASFE Japan.	5.49
2627	1.7.43	238MU, 43GP, India-Vizagapatam, 97MU Ferryside, Newquay, India, Cochin.	8.4.48
2628	1.7.43	238MU, 231ASR Madras.	7.5.48
2629	1.7.43	238MU, 232ASR Columbo. Damaged beyond repair.	18.8.46
2630	1.7.43	238MU, 233 Bombay, 236 Kayts.	23.7.46
2631	20.8.43	Calshot, Felixstowe, Gorleston, Felixstowe, Wick, 62MU, 71ASR Gibraltar, 1110MCU Immingham Store, 238MU. Converted to Target Towing Launch, 238MU Store, Sylt West Germany, 238MU, 1107MCU, 238MU.	12.8.58
2677	20.8.43	Calshot, Grimsby-North Coates, Newlyn, Calshot, AEAF 86GP Gorleston, Lowestoft, Gorleston, 238MU. Converted to Target Towing Launch. 1101MCU Fowey, 238MU, 1113MCU.	———
2678	20.8.43	Calshot, Grimsby-North Coates, Calshot-32-33, AEAF 85GP, Calshot-33, Wick, 62MU, Gibraltar, 238MU. Converted to Target Towing Launch, 1103MCU Felixstowe, 238MU, MEAF.	28.9.59
2679	20.8.43	Calshot, Gorleston, Gt. Yarmouth, Gorleston, Lowestoft, Gorleston, 62MU, 213MU. Converted to Target Towing Launch, 1100MCU Alness.	———
2680	20.8.43	238MU, Bombay-233, Columbo-237, Madras- 231.	7.5.48
2681	13.9.43	238MU, 234ASR, Calcutta-229, Chittagong-228.	7.5.48
2682	20.2.47	238MU, Columbo, China Bay-235, Trincomalee.	20.2.47

RAF BOAT No.	ENTERED SERVICE	BASED AT OR BASE No.	STRUCK OFF CHARGE
2683	25.9.43	238MU, China Bay-235, Calshot, 219ME.	———
2684	10.12.43	238MU, Kayts, Vizagapatam.	23.7.46
2685	15.12.43	Calshot, 2388MU, Vizagapatam.	23.7.46
2686	15.12.43	238MU, Calshot, Vizagapatam.	7.5.48
2687	2.44	Calshot, 238MU, Hamworthy, Calshot, AEAF 85GP, Felixstowe, 62MU, 51EU Bombay, 213MU, Columbo-203.	31.10.46
2688	14.12.43	Calshot, Felixstowe,.Calshot, AEAF 85GP, Felixstowe, 213MU, 238MU, 1107MCU Newhaven, Newquay, Calshot, Pembroke Dock, Converted to Target Towing Launch.	———
2689	2.2.44	Calshot, Felixstowe, Calshot, AEAF 85GP, 238MU. Converted to Target Towing Launch, BAFO Leubeck.	———
2690	2.2.44	Calshot, Felixstowe, Calshot, Felixstowe, 62MU. Converted to remote control Target Launch, 1101MCU Fowey, 238MU. Converted to Target Towing Launch, Pembroke Dock, 1108MCU Blyth.	———
2691	2.2.44	Calshot, Warmwell, Weymouth, Gorleston, Lowestoft, 238MU. Converted to Target Towing Launch, 238MU, FEAF Seletar.	2.5.52
2692	2.2.44	Newquay-238MU, China Bay-235, Chittagong-239.	8.4.48
2693	18.1.44	Calshot 85GP, Calshot-62MU, 84MU, Bombay, Columbo-203.	7.5.48
2694	18.1.44	Calshot, Warmwell, Gorleston, 238MU. Converted to Target Towing Launch.	12.4.50
2695	3.1.44	238MU, Calshot, Hamworthy, Calshot, AEAF 85GP, Calshot, 62MU Heap, 242ASR Bombay.	4.4.49
2696	3.1.44	238MU, Warmwell, Lowestoft, 85GP, 33ASR Calshot, 85MU. Converted to Target Towing Launch, Calshot, BAFO Sylt, West Germany.	8.55
2697	18.2.44	Calshot, Warmwell, Gorleston, Lowestoft, 54ASR, 238MU, 390MU.	———
2698	13.4.43	Calshot, Gosport, Falmouth, AEAF 85GP, Calshot, 62MU Heap, 129 Staging Post, Cocos Islands.	12.5.48
2699	21.4.43	Calshot, Algiers, Italy, 254ME, 207ASRME, Algiers, Piraves.	11.3.48
2700	31.5.43	Calshot, 238MU, 213ASR, Alexandria. Converted to Target Towing Launch.	———
2701	21.6.43	238MU, 213ASRME, 207ME. Wrecked.	13.8.45

RAF BOAT No.	ENTERED SERVICE	BASED AT OR BASE No.	STRUCK OFF CHARGE
2702	28.6.43	238MU, 242 Bombay, 129, Cocos Islands.	7.5.48
2703	4.6.43	Calshot, 238MU, 238ASR Cochin, India.	8.4.48
2704	6.7.43	238MU, Columbo-232/203/237, Madras, Columbo 237.	7.5.48
2705	6.7.43	238MU, 241 ASR, Jiwani.	5.49
2706	6.7.43	Calshot, Gt. Yarmouth. Lost by United States Air Force action.	3.3.44
2707	19.8.43	Calshot, Felixstowe, Gt. Yarmouth, 85M, Calshot, Warmwell, Gorleston, 238MU.	27.5.48
2708	23.8.43	238MU, Iwakuni, Jiwani, Japan, Converted to Target Towing Launch.	———
2709	13.9.43	238MU, Calcutta, Chittagong, Calcutta.	7.5.48
2710	13.9.43	238MU, China Bay, Trinconalee.	7.5.48
2711	12.10.43	238MU, Calcutta, Chittagong.	7.5.48
2712	12.10.43	238MU, Columbo, 237 Jolle.	7.5.48
2713	28.10.43	238MU, China Bay, 210ASR. Converted to Target Towing Launch.	———
2714	5.11.43	238MU, Columbo, Calcutta, Chittagong.	7.5.48
2715	17.11.43	238MU, Columbo, Calcutta.	7.5.48
2716	2.2.44	238MU, China Bay-239, Chittagong-239.	28.2.47
2739	16.3.45	62MU, 238MU. Converted to Target Towing Launch.	12.4.50
2740	16.3.45	62MU. Converted to Target Towing Launch, 238MU, 1110MCU Immingham.	9.6.52
2741	10.6.45	62MU. Converted to Target Towing Launch, Blythe, 1108MCU, 1101 Fowey.	———
2742	10.6.45	62MU, Converted to Target Towing Launch Blythe, 1108MCU.	———
2743	10.11.45	62MU. Converted to Target Towing Launch, 238MU Calshot, 1110MCU Immingham, 238MU, 1113MC, MCTS Mountbatten, 1113MCU. RNS Disposal Mountbatten.	10.62
2744	10.11.45	62MU, 238MU. Converted to Target Towing Launch, 1100MCU Alness. Converted to house boat, burnt.	11.76
2745	18.3.46	238MU. Converted to Target Towing Launch.	———
2746	18.3.46	238MU, 1101MU Fanara.	———

LONG RANGE

The RAF Long Range Rescue Craft or LRRC's were D-Type Fairmiles converted from Royal Navy motor torpedo boats. They were much larger, roomier and more powerful than the Whalebacks and Hants and Dorsets, being 115 ft long by 21½ ft across the beam. Their top speed was 33 knots. Normal cruising at 28 knots gave them a range in the region of 1500 miles.

These boats were powered by four Packard W14 engines derived from the Liberty engine design. Each engine had 12 cylinders arranged in a six-a-side Vee. Engine power was 920 hp at 2000 rpm while cruising and 1250 hp at 2400 rpm flat out. Fuel used was 100 octane petrol.

Forty of these craft were converted for operations in the Far East. They were formed into squadrons of eight and the initial two squadrons left Britain for India in mid-1945. The first had reached Egypt and the second Gibraltar when the atom bombs fell on Japan bringing that war to an end. Some LRRC's remained in the Middle East but the majority returned to home bases. They were roomy and seaworthy, yet were not around long enough to acquire a reputation or a nickname.

RAF BOAT No.	ENTERED SERVICE	BASED AT OR BASE No.	STRUCK OFF CHARGE
001 (MTB 779)	13.7.45	To RAF, 101 ASR/U Calshot, 103 ASR/U Calshot, 102 ASR/U Calshot, Lossiemouth, Calshot. Returned to Admiralty.	20.9.46
002 (MTB 800)	13.7.45	To RAF, 101 ASR/U Calshot, 103 ASR/U Calshot, 102 ASR/U Calshot, Lossiemouth, Calshot. To Admiralty Gillingham.	9.8.46
003 (MTB 5004)	6.12.45	Calshot, 42 ASR/U Newlyn, 24 ASR/U Gorleston, Calshot. To Admiralty.	25.7.46
004 (MTB 5006)	8.45	104 ASR/U Calshot, 42 ASR/U Newlyn, 24 ASR/U Gorleston, Calshot. To Admiralty.	5.10.46
005 (MTB 5011)	26.4.45	101 ASR/U Calshot, 101 ASR/U M.E.D.M.E. at Malta. Converted to Houseboat. Fanara.	9.47
006 (MTB 5012)	30.3.45	101 ASR/U Calshot, 101 ASR/U M.E.D.M.E. at Malta. Converted to Houseboat. Fanara.	9.47
007 (MTB 5014)	7.5.45	101 ASR/U Calshot, 101 ASR/U M.E.D.M.E. at Malta. Converted to Houseboat. Fanara.	9.47
008 (MTB 5016)	26.4.45	101 ASR/U Calshot, 101 ASR/U M.E.D.M.E. at Malta. Offered for disposal.	5.49
009	1.8.45	101 ASR/U Calshot, 103 ASR/U Calshot, 102 ASR/U Lossiemouth, Calshot. To Admiralty.	20.9.46

L.R.R.C. 001 and 003 tied up alongside in Newlyn harbour.

(Photo by M. MacLean from the collection of J. Pritchard)

L.R.R.C. 003 in Newlyn harbour, early in 1946.

(Photo by A. MacLean from the collection of J. Pritchard)

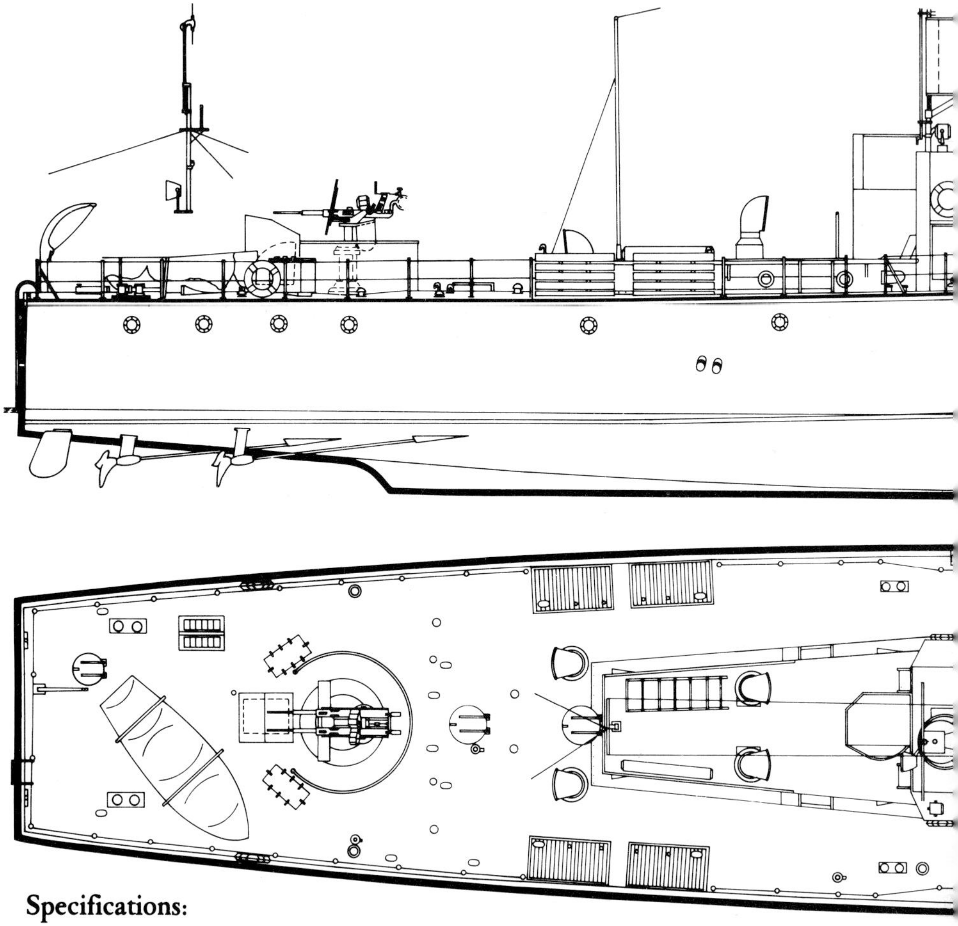

Specifications:

Length: 115ft o.a. Beam: 21½ft. Draught: 5ft.
Power: 4 Packard supercharged petrol engines of 1,250 hp each. 2—12½kW 220v Ford auxiliary generators.
Speed: 33 knots at 2,400 rpm (max.)
28 knots at 2,000 rpm (max continuous.)
Range: 1500 miles plus at 20 knots.
Armament: Twin 20mm Oerlikon cannon aft. 2 X 20mm forward.

FAIRMILE D TYPE 115ft RAF

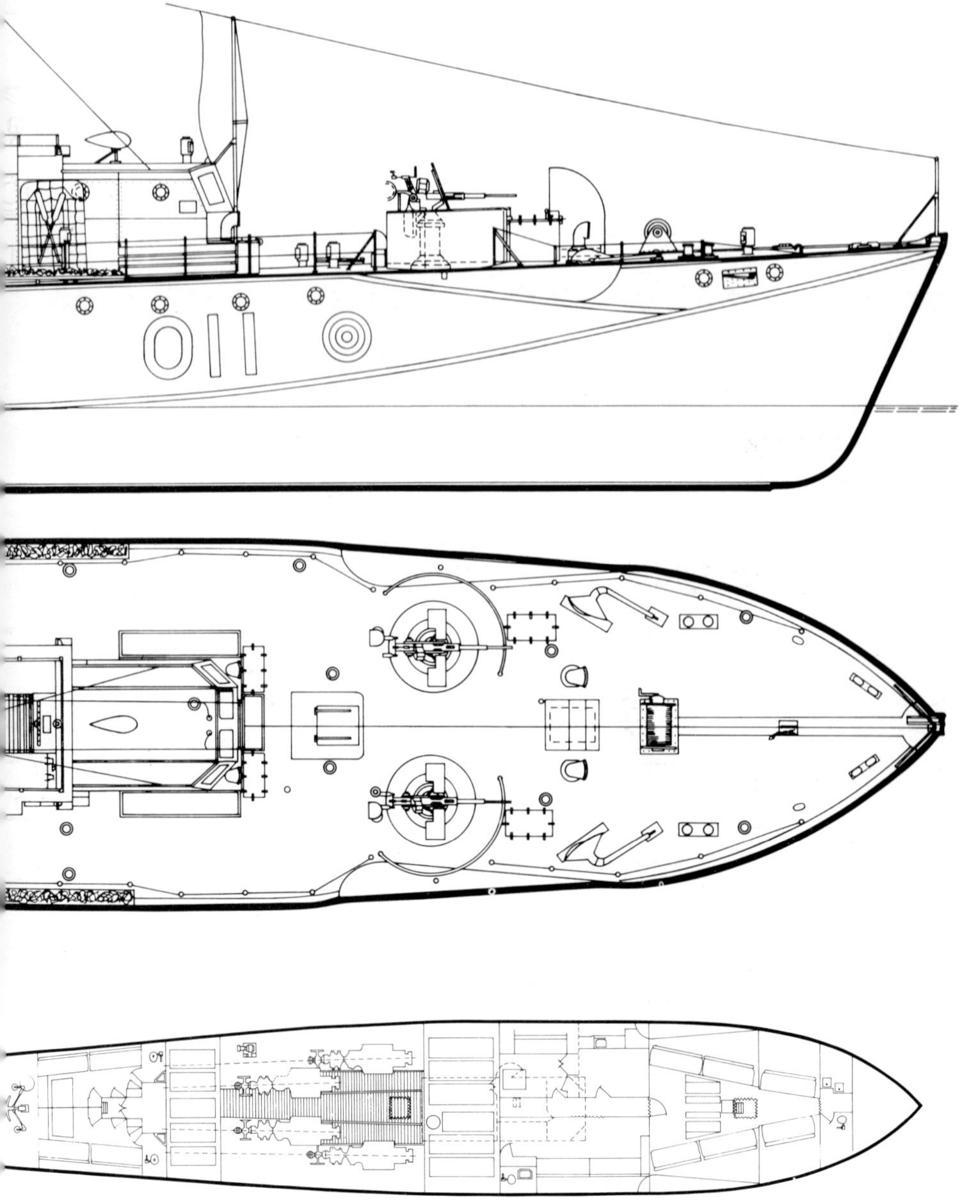

LONG RANGE RESCUE CRAFT

RAF BOAT No.	ENTERED SERVICE	BASED AT OR BASE No.	STRUCK OFF CHARGE
010	3.9.45	102 ASR/U Calshot, Lossiemouth, Calshot. To Admiralty.	24.6.46
011 (MTB 5019)	18.6.45	101 ASR/U Calshot, 101 ASR/U M.E.D.M.E. at Malta. Offered for disposal.	5.49
012 (MTB 5021)	13.7.45	To RAF, 101 ASR/U, 103 ASR/U, 67 ASR/U Bowmore, Calshot, 24 ASR/U Gorleston, Calshot. To Admiralty.	30.9.46
013	15.5.4	Received 101 ASR/U, 102 ASR/U at Gibraltar but returned to UK, Lossiemouth, 102 ASR/U, Calshot. To Admiralty.	20.9.46
014	1.8.45	101 ASR/U, 103 ASR/U, Beccles, 24 ASR/U Gorleston, 40 ASR/U Weymouth, Calshot. To Admiralty.	20.9.46
015	18.6.45	101 ASR/U, 102 ASR/U at Gibraltar but returning to UK, 102 ASR/U Lossiemouth, Calshot. To Admiralty.	24.6.46
016	3.9.45	102 ASR/U Lossiemouth, Calshot. To Admiralty.	24.6.46
017 (MTB 5026)	7.5.45	101 ASR/U, 101 ASR/U M.E.D.M.E. at Malta. Offered for disposal.	5.49
018 (MTB 5028)	18.6.45	101 ASR/U, 103 ASR/U Calshot, 102 ASR/U, Lossiemouth, 102 ASR/U Fraserburgh, Calshot. To Admiralty.	4.11.46
019 (MTB 5029)	1.8.45	101 ASR/U, 104 ASR/U, 67 ASR/U Bowmore, Calshot. To Admiralty.	24.6.46
020 (MTB 650)	26.4.45	101 ASR/U, 101 ASR/U M.E.D.M.E. at Malta. Converted to Houseboat Fanara.	6.47
021 (MTB 664)	26.4.45	101 ASR/U, 101 ASR/U M.E.D.M.E. at Malta. Offered for disposal.	5.49
022 (MTB 673)	26.11.45	Corsewall, Calshot, 213MU. To Admiralty for Timbacraft Ltd pending disposal.	9.8.46
023 (MTB 675)	6.12.45	Calshot. To Admiralty.	18.3.46
024	11.12.45	105 ASR/U Calshot. To Admiralty.	18.3.46
025	11.12.45	105 ASR/U Calshot. To Admiralty.	18.3.46
026	6.12.45	Calshot. To Admiralty.	18.3.46
027	11.12.45	105 ASR/U. To Admiralty.	18.3.46
028 (MTB 680)	12.7.45	101 ASR/U. 103 ASR/U, 67 ASR/U Bowmore, Calshot. To Admiralty.	18.3.46
029 (MTB 682)	1.8.45	101 ASR/U, 103 ASR/U, Lossiemouth, Calshot. To Admiralty.	18.3.46
030	1.8.45	101 ASR/U, 102 ASR/U Calshot, Lossiemouth, Calshot. To Admiralty.	18.3.46

RAF BOAT No.	ENTERED SERVICE	BASED AT OR BASE No.	STRUCK OFF CHARGE
031 (MTB 684)	16.5.45	101 ASR/U Calshot, 102 ASR/U Calshot, 101 ASR/U M.E.D.M.E. at Malta. Converted to Houseboat Fanara.	1.47
032 (MTB 687)	18.6.45	101 ASR/U Calshot, 102 ASR/U Calshot, 104 ASR/U Calshot, Attached to Dorset Yacht Co. for shipping and repairs, Calshot. To Admiralty.	18.3.46
033 (MTB 689)	8.45	104 ASR/U Calshot, Calshot. To Admiralty.	24.6.46
034	1.8.45	101 ASR/U, 103 ASR/U, Calshot. To Admiralty.	18.3.46
035	9.45	104 ASR/U, Calshot. To Admiralty.	24.6.46
036	16.5.45	101 ASR/U, 102 ASR/U, 101 ASR/U M.E.D.M.E. at Malta. Converted to Houseboat Fanara.	29.4.47
037	11.12.45	105 ASR/U Calshot. To Admiralty.	18.3.46
038	11.12.45	104 ASR/U Calshot. To Admiralty.	20.9.46
039 (MTB 695)	9.45	104 ASR/U Calshot, Calshot. To Admiralty.	24.7.46
040 (MTB 744)	18.6.45	101 ASR/U, 102 ASR/U, Lossiemouth, 24 ASR/U Beccles, 24 ASR/U Gorleston, 40 ASR/U Weymouth, Calshot. To Admiralty.	20.9.46

The majority of these RAF Long Range Rescue Craft were returned to the Royal Navy. Some were converted to houseboats and used as married quarters, while others were disposed of locally.

The following table is a brief comparison of the launches featured in this book.

Type	**63ft HSL 'Whaleback'**	**68ft HSL Hants & Dorset**	**115ft D Type 'Fairmile' Long Range**
Builder	British Power Boat Company	British Power Boat Company	(Ex Royal Navy MotorTorpedo Boats)
Construction	Double diagonal mahogany planked hull.	Double diagonal mahogany planked hull.	Semi hard chine. Double diagonal hull & deck. Sheathed in copper below the water line.
Length — overall	63ft	68ft	115ft
Beam	17ft 2in	17ft 1in	21ft 6in
Draft	3ft 10in	4ft 3in	5ft 6in

Engines — Main	3 x Napier Sea Lion, 500 BHP	3 x Napier Sea Lion, 500 BHP	4 x Packard W14, 1250 BHP
Fuel — Type	Petrol	Petrol	Petrol
Quantity	1,200 gallons	1,700 gallons	8,800 gallons
Consumption	109 galls/hr		
Speed — Maximum	35 knots	28.5 knots	33 knots
Cruising	28 knots	23.5 knots	20 knots
Displacement		32 tons	
Propellers — Type	Nickel alum bronze (LH)	Nickel alum bronze (LH)	4 x 3-bladed (RH) (RH)
Pitch	—	19½in	45½in
Diameter	—	22in	36in

Like the Whalebacks and the Hants and Dorsets, the LRRC's are still seen around the coast, either up some creek or on a mudflat, overshadowed by more exciting present day craft. But once they represented the latest in high speed motor boating, their paintwork shone and they were manned by proud crews. They helped in a worthwhile cause and should not be forgotten.

After 1st April 1986 the RAF Marine Branch will have been disbanded and the RAF sailors will be no more. For the first time in sixty years the Royal Air Force will be without its own boats and without airmen eager to man them.